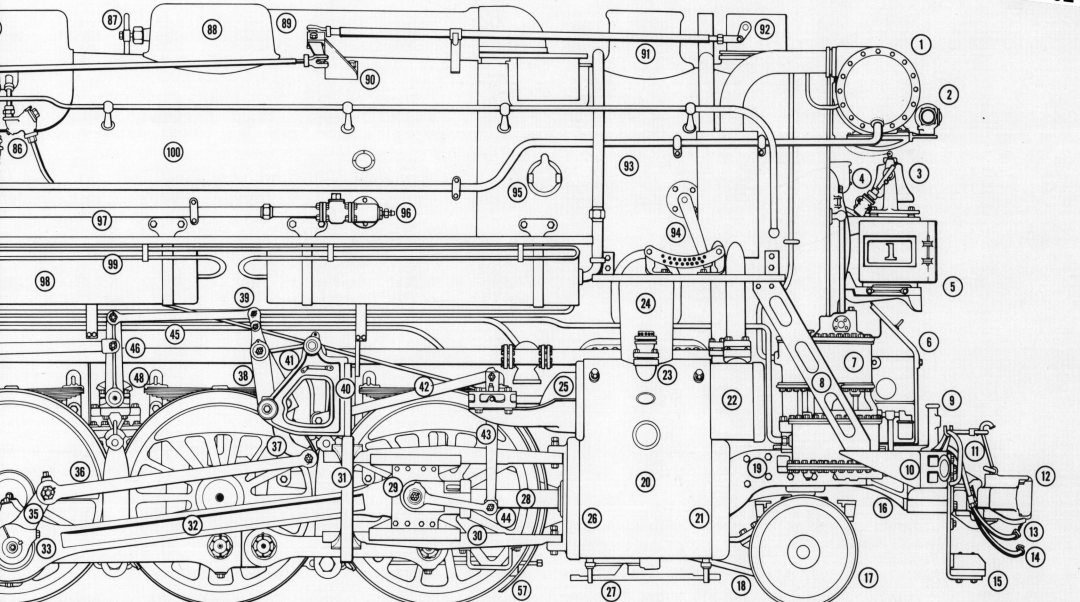

52. Driving axle
53. Driver
54. Counterweight
55. Brake hanger
56. Brake shoe
57. Sand pipe
58. Furnace bearer
59. Firebox
60. Mud ring
61. Ashpan
62. Ashpan hopper
63. Trailing truck
64. Trailing truck spring rigging
65. Pads and centering devices
66. Booster
67. Booster exhaust pipe
68. Injector

69. Injector feed pipe
70. Drawbar
71. Drawgear radial buffer
72. Cab handrail
73. Stoker
74. Cab
75. Ventilator
76. Steam turret
77. Turbo-generator
78. Safety valve
79. Whistle
80. Auxiliary dome
81. Throttle rod
82. Handrail
83. Running board
84. Oil separator
85. Sandbox

86. Sander valves
87. Dome shut-off valve
88. Steam dome
89. Dry pipe
90. Throttle bell crank
91. Smokestack
92. Throttle valve
93. Smokebox
94. Variable exhaust quadrant
95. Cinder cleaning hole
96. Check valve
97. Injector delivery pipe
98. Main (air) reservoir
99. Radiator
100. Boiler

SUPERPOWER

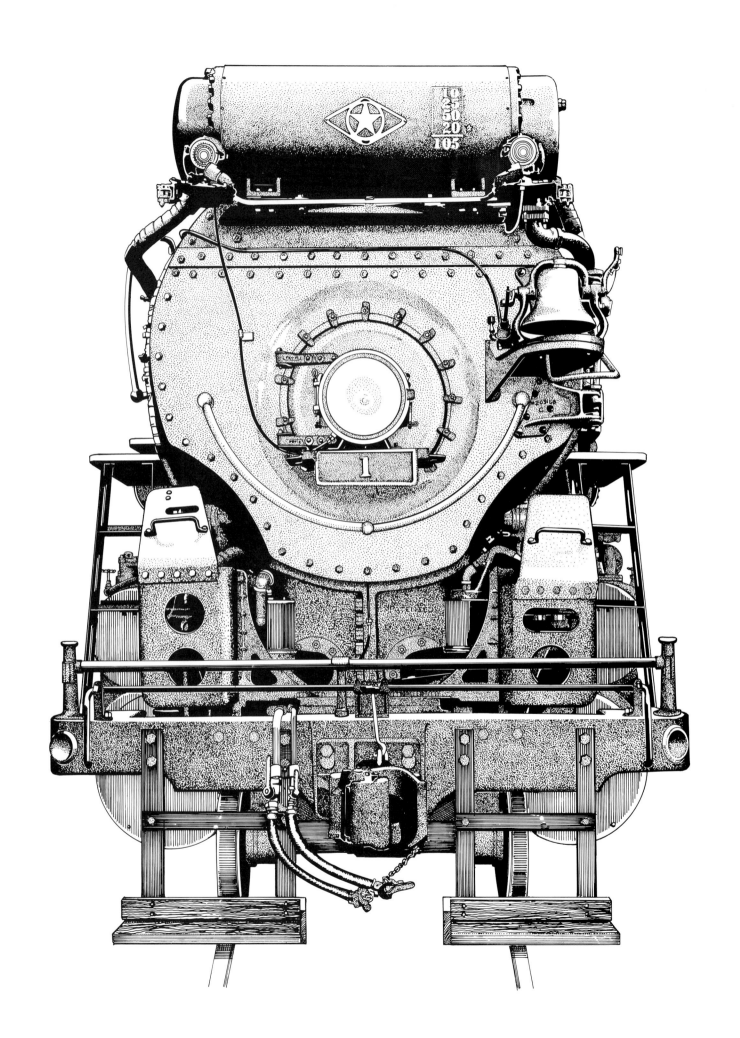

SUPERPOWER

The Making of a Steam Locomotive

written and illustrated by

DAVID WEITZMAN

DAVID R. GODINE · *Publisher* · *Boston*

First edition published in 1987 by
DAVID R. GODINE, PUBLISHER, INC.
Box 9103
Lincoln, Massachusetts 01773

LC: 86-46255
ISBN: 0-87923-671-X

Third printing, 1995
Printed in Spain by Imago Publishing, Ltd.

For Brooks

Contents

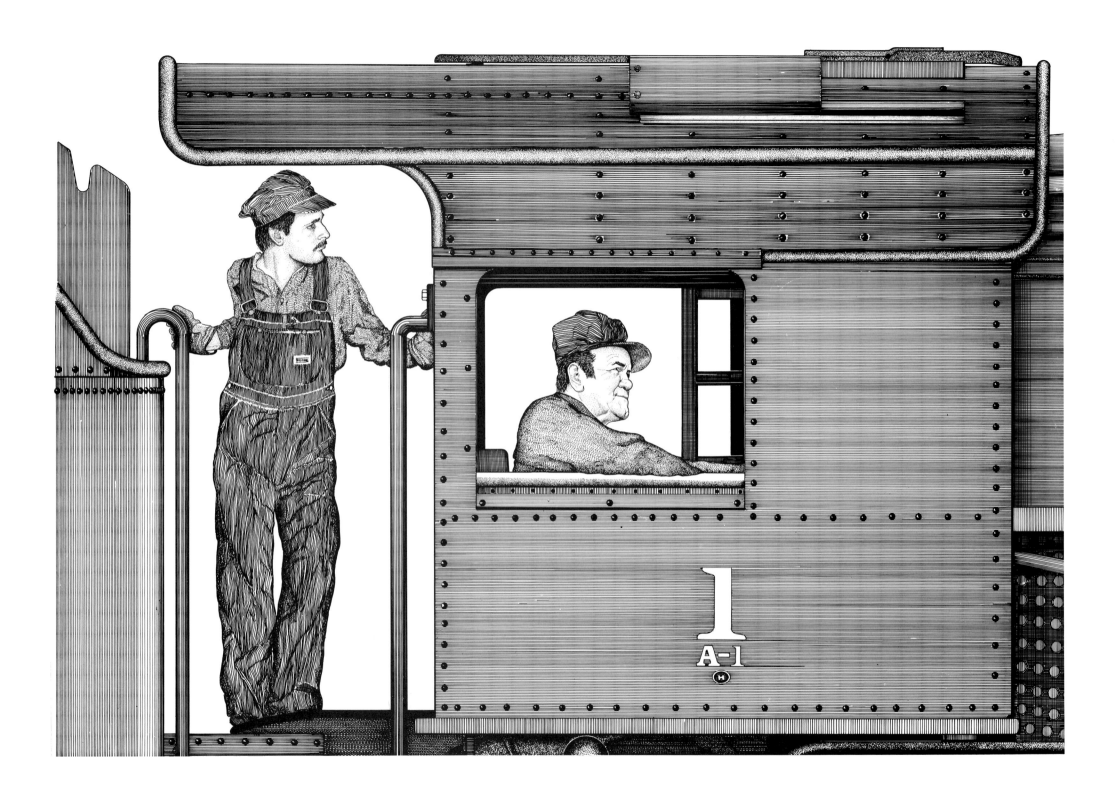

Prologue

APRIL 14, 1925. It is morning at the Boston & Albany's Selkirk, New York, yard, and thousands of freight cars line storage tracks to the horizons. Strings of dusty black hoppers heaped with coal dug from Pennsylvania and West Virginia mines. Tank cars black with dripping crude from Texas and Oklahoma oilfields. Red cabooses. Gondolas loaded with bulging burlap-wrapped bales of Arkansas cotton. Wooden refrigerator and boxcars trumpeting *Curtiss Baby Ruth, Ralston Purina, Edelweiss Brew* and *Yakima Valley Apples* in bright orange, blue, red, yellow and white. Green express cars. Tuscan red boxcars packed with machinery, cook stoves and North Dakota wheat. Low flatcars stacked with Oregon lumber and steel pipe from Ohio mills and lined with apple-red harvesters from Illinois.

A haze hangs overhead fed by columns of coal smoke and soft plumes of steam rising from the locomotives shuttling back and forth between the roundhouse, coaling towers, water plugs, shops and waiting trains. Small six- and eight-wheel switch engines—yard goats they're called—dart through a confusion of switches picking up a string of flatcars here, setting out cattle cars there, adding on to trains that will stretch almost a mile between locomotive and caboose.

It is now 10:57 A.M., and engine No. 190 steams out of the yard, eastbound to Springfield, Massachusetts, with a train of forty-six freight cars. Union Pacific. Santa Fe. New York Central. Nickel Plate. Baltimore & Ohio. Pennsylvania. Chicago & North Western. Cotton Belt. Missouri Pacific. Rock Island. Delaware, Lackawanna & Western. Erie. Great Northern. Wabash. Boston & Maine. Engine No. 190 is a 2-8-2 (two small wheels up front in the lead truck, eight big drive wheels and a two-wheel trailing truck under the cab). It's a Mikado, or just "Mike" to railroad men. And it's a powerful locomotive, the most powerful in its class. Even so, No. 190 might take eight to ten hours to go a hundred miles with its long, slow "drag" freight.

Clanging bells, staccato whistle blasts, chuffing engines, lowing cattle, cars slamming to couple: all punctuate the incessant rumble of work. Signal lamps flash red, yellow, green. Striped semaphores, perched high up on signal bridges like flocks of cranes, swing up and then drop silently, starting, slowing, stopping, holding traffic. Brake shoes squeal against steel wheels. Switchmen and brakemen swing their lanterns in graceful arcs, signaling come ahead . . . stop . . . come back . . . slow . . . hold it. It's like this day and night at Selkirk yard and hundreds of similar yards around the country, strings of cars lengthening and dividing, trains arriving and departing every few minutes.

Drifting smoke and steam, caught in a sudden gust, sweep down, engulfing figures standing beside their locomotives. Out around the engine yard the ritual of firing, watering, coaling and pampering the soot-black giants goes on. Men and machines prepare for a day's hard work together. Crewmen stand around their locomotives, steam swirling about their legs, going over their train orders, telling jokes, catching up on the latest with someone who has one of those newfangled radios. Engineers consult gold pocket watches on long, slender chains—timepieces that err only a few seconds a month. Firemen clamber up onto the running boards, checking a sanding valve high up on the boiler or polishing a headlight lens. An engineer pokes the long tapered spout of his oil can toward a thirsty oil cellar between spoked drive wheels. He pauses to wipe up a drip, carefully, and to give a rod, crank or some other cherished part a polish with his oil rag.

The great engines are regarded with respect, even love, by their crews. They are partners. The men enjoy being in the presence of these gentle giants, thrill to control such power, to be part of railroading.

The time, according to the tower clock and a hundred gold Waltham watches, is now 11:05 A.M. There's a stir in the yard, the sudden awareness of some presence. Brakemen signal their switchers to a halt and climb up on the running board for a look. Engine crews lean intently from their cab windows. Shopmen stand in the roundhouse doorways, watching. Track workers stop, look up, rest on their shovels and picks, frozen in a work tableau. Up in the brick tower shirt-sleeved signal operators and dispatchers gather at the large windows overlooking the bustling yard. All eyes are on a shining black hulk of a locomotive steaming off the turntable and out into the yard, threading its way effortlessly, silently, through the labyrinth of switches, heading for a waiting string of freight cars.

There have been larger locomotives through these yards, but there's something remarkable about this one. From its proportions—the massive firebox and boiler, the huge cylinders and eight white-rimmed drivers—comes an impression of awesome power. There's power in the ease

with which it moves, silver rods and valve gear stroking back and forth in the precise tempo of some immense clockwork. Painted on the cab, instead of the usual three- or four-digit number, is simply A-1. And under the ash-gray firebox, radiating heat waves in the morning sunlight, is a peculiar trailing truck with four wheels. That makes it a 2-8-4, a new wheel arrangement not seen in this yard before. Even now, coasting slowly, smoothly through the yard, the A-1 shows a penchant for speed.

The locomotive stops. The engineer looks back from high up in his cab window at the brakeman standing behind the long tender, anticipating his signal. The brakeman swings his arm in a circle, signalling "move back." The power reverse shifts the valve gear into reverse position. And A-1 backs through a lineup of switches and couples to the waiting string of cars, sending a crunching jolt back through the whole train.

Engineer Bob Keller spins the power reverse wheel, returning the valve gear to the forward position. He lets his locomotive creep forward just enough to make sure the hitch has been made. The brakeman hooks up the air hoses between the tender and first car. Meanwhile, a couple of "car knockers" walk the full length of the train checking that the brakes on each car are working. The engineer and fireman sit at their windows looking expectantly at one of the dozens of semaphore signals ahead. Both are feeling a kind of quiet excitement as they wait, listening to the roaring coal fire between them, the growl of the stoker and the whine of the steam turbo generator.

Bob looks over the array of red-handled valves and glass-faced gages in front of him on the back of the boiler. The pointers on the booster-engine, feedwater-heater, boiler-pressure and stoker-engine gages tell him the locomotive is ready. Waiting for the report of the car knockers, in the heat of the cab, engineer and fireman sit back in their black leather seats. They look out the cab windows, their heartbeats picking up on the measured, throbbing *thrupp shhhr . . . thrupp shhhr . . . thrupp shhhr thrupp shhhr* of the air compressors pumping up the train line.

The A-1 has just come from the Lima Locomotive Works, and will be making trial runs on the line between Albany and Boston. Today's run is a test, so Bob's train orders tell him that he is to run without speed restrictions. He was chosen especially for this run. Bob's been at the throttle for over thirty years now, and has trained many a young fireman like Jack Mustanen, who sits across the cab from him today. During idle moments like these, Bob's mind wanders back over the years of railroading, to his first job as a fireman. He's been in the cab of some big locomotives, but nothing like this morning. Instinctively, his experienced eyes scan the gages again.

"Fifty-four cars all working," the car knocker shouts from the trackside below the cab, interrupting Bob's reverie. Bob looks to his conductor. Highball. "High green," the fireman calls from the other side of the cab. The semaphore signal has snapped to the upright position as if saluting the departing train. Bob digs an aged gold watch out of his overall bib with the kind of flourish that goes with big gold pocket watches on long chains, and springs open the delicately engraved top: 11:44.

"Let's go!" Bob reaches up for the whistle cord and pulls off two short blasts, acknowledging the tower's signal and clearing the train. Automatic bell ringer: on. Steam pressure: 240 psi (pounds per square inch). He sets the valve gear fully forward. Then he opens the sander valve, which releases sand from the large dome on top of the boiler, sending it down through pipes and onto the rails just in front of the first pair of drivers. Bob opens the cylinder cocks to blow out any condensation. Then he cracks the throttle, throwing steam into the cylinders. A-1 responds immediately. Sure footed—no slipping drivers even with this heavy load—the locomotive creeps forward under

explosive gusts of smoke and steam. Underneath the cab, in the new four-wheel trailing truck, pulses a small steam engine, a booster, which adds to the locomotive's starting-up horsepower. (Bob will turn it off when he's under way and then use it again going into steep grades.) Engine and train are rolling.

"Bad iron," the fireman calls out. He's spotted a switch in the long lineup ahead that's not set right. Bob eases back on the throttle letting A-1 coast, and pulls off four short whistle blasts, his call to the tower for signals. Presently the points of the errant switch slide over, the target on the switchstand shows green and the fireman acknowledges, "OK on the iron." Bob cracks the throttle again and A-1 clatters through the jumble of switches, crossing, intersecting, crossing back.

At the throat of the yard scores of tracks narrow to just a few leading out to the "high iron," the railroad highway. Picking up speed, A-1 makes its way down the industrial corridor, a man-made canyon of red brick factories with sawtooth roofs and monumental power stations with towering smokestacks. Along the tracks are great pyramids of coal and crushed rock. A forest of cranes pierces the sky etched by a web of electric power lines. Steam shovels bite into mountains of ore. Signal towers and bridges, storage tanks, green gas holders, refinery towers rise up from the industrial landscape. Strings of loading and unloading freight cars line sidings.

A-1's exhaust cracks sharply off the hard brick, concrete and wired-glass walls of the canyon, echoing back to the crew from every direction. After a few miles, the factories become sparse, separated by tall-grass marshes. Then the corridor opens up into freshly plowed and seeded fields, pasture and scattered woods. At the insistence of Bob's gloved hand, A-1 begins stepping out. Each pull on the throttle brings a surge of power from the locomotive and approving nods between fireman and engineer. Suddenly the roll of the wheels on the rails changes to a hollow drumbeat as locomotive and crew leap out into space, suspended over the churning waters of the Hudson River, caught in a zigzag blur of silvery steel lattice girders.

This run is familiar to the engineer; it's been his for years. After leaving the yards and the corridor, A-1 heads out across the broad Hudson River Valley, through forest tangles of birch, maple, beech, hemlock and mountain laurel, past whitewashed clapboard and steepled towns. When the forest gives way, it is to rolling meadows, mill ponds dotted with ducks, gambrel-roofed red barns on fieldstone foundations and clusters of round-domed silos. Horses at the split-rail fences, surprised by the steaming giant, gallop away, white eyes wide, feathery manes flying to the wind.

The tracks climb into the rimpled foothills of the Berkshires, gradually at first, then getting steeper and steeper. Locomotive, engineer and fireman plunge together into the dark of long tunnels and soar over deep rifts on towering trestles. The exhaust reverberates off the vertical walls of tilted shale at the track's edge. Fainter echoes return from the distant bluffs topped by stands of white pine. High up in the cab, Jack's and Bob's eyes are dazzled by the midday sun caught flickering in the rivulets and rills of spring snowmelt flashing beneath the wheels.

Here, instead of creosote and coal smoke, the cool air rushing past the cab windows brings sweet smells of rainwashed forest, spring wildflowers, moss and damp earth. The scene is enthralling, especially to the young fireman born and raised in the city's industrial heart. He likes the rush of clean air on his face, in his hair, lets it fill his mouth like cold, fresh gulps of mountain water.

Jack Mustanen has been firing locomotives for five years now, since he was sixteen and first rode with his uncle. (Jack, by the way, is what his friends call him. At home he's Jens, the name his immigrant parents brought with

them from their home in Finland.) This has never been an easy run; just now, as Jack watches out his window, the tracks have begun climbing into the twisting turns and steep grades of the Berkshire Hills. It's a test of any locomotive and crew, at best a slow grind at ten or fifteen miles an hour. Trying to go any faster than that, the engineer risks running out of steam because the conventional firebox and boiler cannot make enough. But the A-1 has been steaming along at thirty, and ahead is another clear block.

"High green," Jack shouts over the roar of the fire and rumbling stoker. One of the fireman's responsibilities is to watch for and call all signals.

"High green," Bob confirms. Thirty, thirty-five, thirty-eight, A-1 reaches forty miles per hour on a long flat stretch. Bob checks his gages again; boiler pressure is rock steady at 240. Jack spots a grade crossing ahead. "Gates are down," he calls to the engineer. Bob acknowledges his fireman's call and grabs overhead for the whistle cord, sounding two long blasts, a short and then a long. Jack waves to the crossing gate tender. Bob raises a gloved hand to the wildly waving children standing atop hay bales in a waiting truck.

This is Chatham. The train speeds by the ochre and brown freight depot and swings alongside High Street, past white shuttered houses with front porches facing onto the tracks. Past the gray-haired woman in a pink gingham dress who watches and waits in her porch swing for the train's passing each morning. Past a little boulder-filled creek meandering alongside the roadbed. Past a red-brick warehouse and a fieldstone forge where a leather-aproned smith stands in the doorway, hammer in hand, aglow with the flickering yellow forge fire. In another moment the town is gone and A-1 plunges into the green dark of the pine forest.

Inside the cab of a locomotive steaming along at forty

miles an hour, all of the senses are touched. You can feel the heat of tons of red-hot coal radiating from the steel firebox. The smells are damp coal, coal smoke, steam, oil, grease and hot machinery. The noise is superb, as breathtaking and big as the machine itself—a raging firestorm, two-hundred tons of steel rolling on steel, hissing steam, puffing blasts of exhaust, whistle shrieking at grade crossings, wheels clattering over switch frogs and crossings, tender and train rumbling behind. It's everything Jack and Bob want it to be. They chose to be enginemen in their youth; the noise is part of the thrill. "It's like having your own thunder," Bob once told his grandchildren.

Jack appears a bit uncomfortable today. But whatever is bothering him, his nature is not to dwell on it. He's busy spotting signals, grade crossings and switches for the engineer whose vision is cut off by the boiler, especially on long, sweeping curves to the left. And on wide, open turns he can look back over the whole train to make sure there are no blazers, the long trail of smoke from a "hot box," or overheated wheel bearing. "All black," he calls out to the engineer. He keeps a vigilant eye on the water glass, turning on the injector to keep water at a safe level in the boiler.

Most important, there's the fire to look after. Jack is responsible for the 240 pounds of steam pressure Bob counts on to climb through these mountains. The fireman has to make sure there's always enough coal on the grates. Jack controls the blower that increases the draft in the firebox whenever the locomotive is traveling too slowly to draw enough air up through the grate. Right now, he's peering up at A-1's stubby smokestack for clues to how well his fire is doing. There's only white steam and light gray smoke bursting out of the stack in time to the exhaust strokes of the pistons. That's the way he wants it. A locomotive billowing dense black smoke adds excitement to train robbery scenes at the nickelodeon, all right, but

to the fireman it's a sign of poor combustion and a warning that he's wasting coal. (It's also a sign to the railroad's smoke inspector who would remind offenders of the operating department's Rule 942: "Firemen must fire their engines, as far as possible, in such a manner as to avoid the emission of black smoke.") Still, with all this to do, Jack looks uneasy, his powerful arms hanging awkwardly at his sides. Bob has noticed and, realizing his partner's predicament, begins kidding him about it. "What's the matter, young man, you lose your coal scoop?"

In fact this is the first time Jack has been in the cab of a locomotive without a coal scoop in his hand. Behind him, in the tender, are eighteen tons of coal, and it has always been his job to get it into the firebox a heavy scoopful at a time.

Jack used to open the fire door frequently to hurl scoopful after scoopful after scoopful into the red-orange inferno. He fired in a prescribed pattern—a scoopful into the far left corner of the firebox, the next scoop into the near right corner, another to the far right-hand corner, another to the near left and then a scoopful to each side. All the time he had to keep the bed of coal about four inches deep all over. And he had to work fast. The draft through the open fire door blew big holes in his fire and cut down on the heat for steam.

It was hard work. The temperature at the fire door is about 2,200 degrees Fahrenheit. When a big engine works really hard, so does the fireman. Jack would shovel over two tons—four to five thousand pounds of coal—scoop by scoop, into the firebox every hour, his face glistening with sweat, watching his gages through salt-stung eyes. Sometimes the head brakeman, who sat behind Jack in the cab, would spell the fireman so he could rest a few minutes. And Jack remembers all too well those New England winter mornings when he would be shoveling and the brakeman would be back in the tender breaking up the

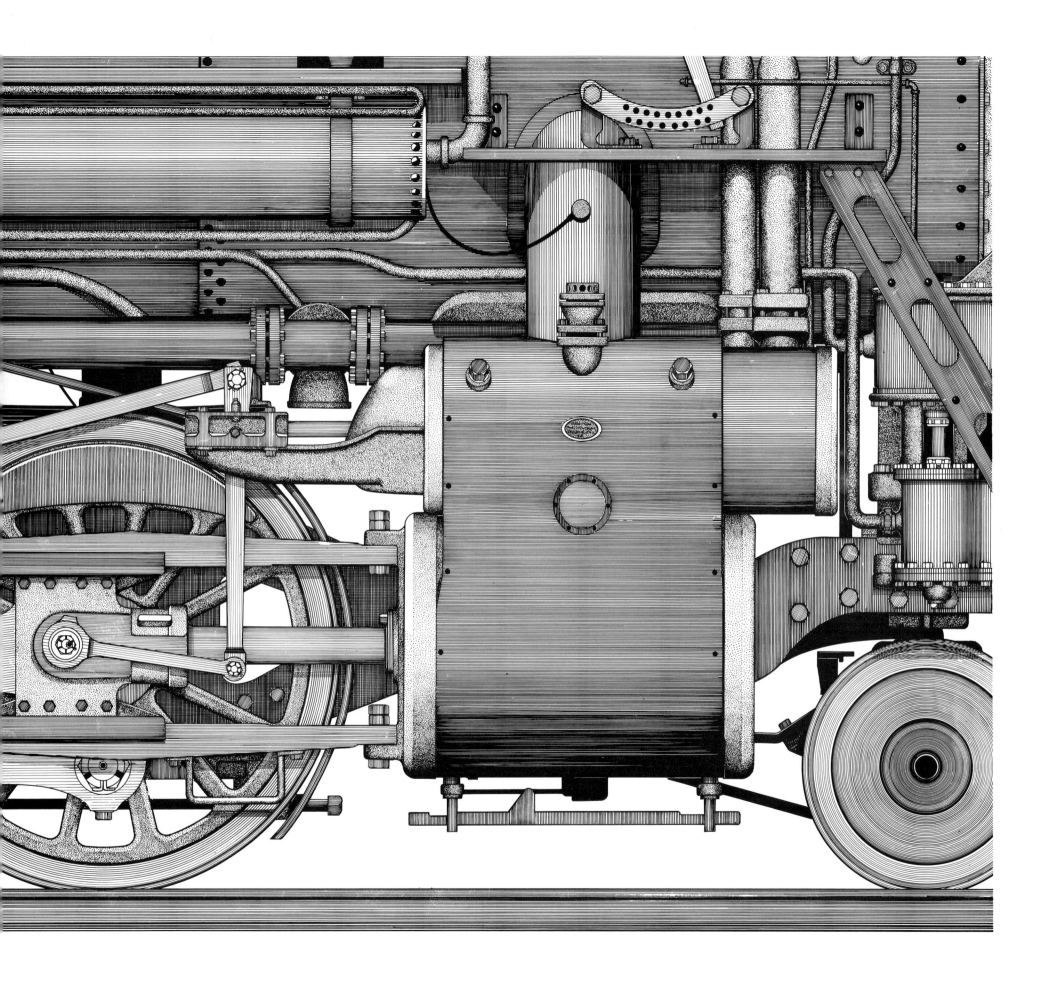

frozen mountain of coal with a pick. That's what it was usually like. But today the A-1 is firing itself.

A-1's firebox is the largest ever built for this class of locomotive. The grate—the cast-steel firebox floor that holds the fire and lets air up through the burning coal—is 100 square feet, about the size of a bedroom. The idea of the bigger firebox and boiler is to boil more water faster, make more steam at a higher pressure and give the A-1 its impressive horsepower. The problem is that a firebox this large makes it impossible for a fireman or even a fireman plus a head brakeman to shovel coal fast enough. Right now, at forty miles per hour, A-1 is burning over three tons—six thousand pounds—of coal every hour. Working at its hardest, this engine could burn as much as seven thousand pounds an hour.

The automatic stoker was the answer. It crushes the coal into small chunks, carries it in a long worm screw from the tender to the firebox and then spreads it evenly over the grate with mechanical shovels. The stoker moves more coal than Jack could, moves it faster, and—though at first he was not wanting to admit it—makes a more uniform, hotter fire.

Jack instinctively reaches for his scoop now and then. The work and the rhythm of the shoveling has always been part of what he does. But gone from the cab today is the cadence of the swinging scoop scraping the coal off the steel cab deck, the ring of the poker and fire hoe against the fire doors, the ritual of the fireman's work. In its place, the steady grumble of the stoker.

So Jack's in his seat more than usual. In front of him are four new valves controlling the speed of the worm screw and the mechanical shovels. "Green eye," he calls out; Bob confirms the clear block signal. Engineer and fireman ride their iron horse past gently rolling hills of rich black soil, down into deep cuts and up along embankments level with tree tops and roofs. Sometimes the trackside

falls away into a steep ravine. Sometimes the crew find themselves under a canopy of maples and birch, the thicket of trunks spreading upward into a lacework of small branches above their heads.

Bob looks at the speed indicator. A-1 has been pounding steadily along at an astonishing forty miles an hour. But neither man really needs a gage. The rush of wind in their faces tells them that this locomotive is moving! "Like ridin' the wind," Bob says to himself. "What's that?" Jack asks. "Nothin', nothin' at all," Bob answers, passing off the remark with a shrug, glad his fireman hadn't heard it, a bit embarrassed by the notion. "Like ridin' the wind."

Bob whistles for the children standing beside the track, awaiting the locomotive in an excited, hesitant way that is something between curiosity and fear. At the last moment the tow-headed brother and sister fling themselves down on the grassy slope of the embankment giggling, waving, fearful of being caught up and carried away in the flying giant's rushing wind.

The A-1 arrives at East Chatham and this is where the real test begins. From here to Canaan will come the steepest grades and tightest curves of the run. In anticipation, Jack checks his fire and, adjusting the stoker throttle valve, sends more coal into the firebox. Slowing only slightly, the giant climbs into the first steep pitch, leaning into the banked turn, the bark of the exhaust becoming more insistent, reverberating off the sandstone bluffs. Jack inspects his fire again, making sure the coal is spread evenly over the grates. Into the white heat of the firebox the stoker sprays a hail of crushed coal which ignites and burns before falling to the firebed. Bob glances again at the steam gage; "Good work, Jack. Give me all the steam you've got." He eases the throttle all the way out, feeling the surge of power through his whole body. A-1 climbs up the tortuous grades like no other locomotive's done before.

"Double iron ahead," Jack announces. Another track

turns off the high iron and runs in a blur of ties and silvery ribbons along with the engine. "Train ahead," Bob calls over to his fireman, silhouetted in the bright rectangle of the cab window. Looking ahead, past the long boiler, Bob sees a caboose, the end of another train. He thinks a moment, a puzzled expression crosses his face, and he pulls out his big gold watch. One o'clock. It has been an hour and twenty minutes since they left Selkirk yard. Then it dawns on him. "Jack, that's 190. The only thing it could be!" A shiver of excitement runs through the young fireman.

Number 190 had been switched to the outside track at East Chatham junction and stands waiting for the faster train to pass. Like one of those all too familiar algebra problems, A-1, which had left the yard almost an hour later, is now overtaking the earlier train. Jack checks his fire and gages once again. Steam pressure: 240. Speed: 38. Water glass: a little low. He turns on the feed pumps and injector sending hundreds of gallons of water into the boiler (preheated to over two-hundred degrees in the feedwater heater mounted just in front of the stack). The stoker is on. The distance between the two trains closes faster and faster.

From inside the caboose a brakeman emerges onto the back platform to watch A-1's approach. (What a picture that must be, Jack thinks to himself.) He waves to the fireman and in another instant is gone. Boxcars, gondolas, hoppers, reefers, tank cars separated by flashes of light fly by only inches from the engineer's window in a phantasmagoria of colors, letters, numbers and flickering shapes. The roar of the two trains is almost painful, every sound amplified in the swaying, rolling cab, getting louder still as A-1 comes up on No. 190. A-1's cab is now alongside 190's tender and then the cab windows are side-by-side. Time and motion seem suspended as Bob and Jack look across into the other cab. It is only an instant. The fireman, shovel in hand, looking up from his fire door and the

engineer at his window are caught in a snapshot with the click of a Kodak shutter and disappear. In another click of the shutter No. 190 is left behind and, with a flash, the cab brightens and the trackside forests reappear. Bob grasps the knotted whistle cord and A-1 shrieks a long, hearty salute to 190's crew. He calls out the clear switch and block ahead, and brings the throttle out to the last notch.

When the 2-8-4 steams past the yard limit sign and into North Adams Junction it is ten minutes ahead of No. 190. The engineer brings his engine and long train to a smooth, gliding stop. A switcher chugs up alongside the A-1 ready to shunt cars out into the yard. The brakeman, standing up front on the footboards, awaits instructions from the conductor now making his way hurriedly toward the engines from the caboose. From here cars will be sent in all directions, to Boston, to White River Junction, to Wilmington, Woburn and Portland.

Bob draws out his watch, with a bit more ceremony this time, holds its face to the light of the cab window and smiles. It is 2:02 P.M.. The A-1, with its longer, heavier train, has taken almost an hour off what had been a four-hour run.

Bob puts his watch back in his pocket, makes a last check of his valves and gages, grabs the clipboard with his train orders and gathers up his jacket and satchel. Then he climbs down the cab steps and walks a few steps away. He takes off his cap and glasses, and wipes the grime from his wrinkled face with his blue-checked bandanna. After polishing the dust off his glasses, he places them back on his nose, hooks the bows behind his ears and takes a good, long look at this locomotive he's been driving.

Jack joins his partner and stands at his side, quietly, looking over their engine. Running through their minds are a lot of unspoken things. All the locomotives they've known. And how only a month or so ago the laboring Mike they passed up back there seemed the best ever. Their eyes are filled with the new 2-8-4. "That's as good as they make 'em, son." "Yep," the fireman agrees, "it's a good one, all right." It is the finest locomotive on any railroad, they are sure of that, and they had been chosen as crew.

"I'll take 'er over to the ash pits now." The arrival of the hostler, the engineer responsible for locomotives in the yard, interrupts quiet reflections on a sunny spring afternoon. The brakeman uncouples the engine from its train as the hostler climbs up into the cab. A-1's brake shoes fall away from the drivers with a whoosh of air and the locomotive slowly steams away toward the roundhouse.

PutputHONKputputputputHONKHONKputputputput. "What in Hades?" A boxy yellow switch engine, half the length of A-1's tender rattles up behind the engineer and fireman, sending them scurrying from the tracks.

"It's called a 'diesel.' Got one of those internal combustion engines. Runs on oil," Bob says. "Read in a magazine that diesels are someday goin' to replace steam locomotives." Jack shakes his head in disbelief and laughs as he watches the comical little engine pull up alongside the 2-8-4 towering above it. "Suuure don't mount to much."

Bob grabs his young fireman roughly by the scruff of the neck, squashes his rumpled cap down over his eyes and gives him a good shaking. Shoulder to shoulder, laughing, each with a black tin lunch pail in his hand and heavy work gloves and jacket tucked under his arm, they amble down the track toward the engine house.

On their return trip westward, Bob and Jack and the A-1 haul eighty-nine empty refrigerator cars back to Selkirk—in these days an unthinkable tonnage for such a mountainous run. In the next few months, A-1 breaks other records on other runs, going alone long distances that had required several engine changes along the way. The Boston & Albany immediately places an order for fifty-five of the new 2-8-4s, dubbing them the "Berkshires," for the rough terrain where they had proved and distinguished themselves. The Illinois Central orders another fifty, and over the next twenty-five years more than six hundred Berkshires are built for freight and passenger service.

The A-1 was a line drawn between the old and the new. "Those Berkshires," a retired B&A engineer still recalls, "revolutionized the railroading we had known." What follows is the story of the men who designed and built the first Berkshire.

·1·

Beginnings

BOUND FOR Lima, Ohio, the silver rails of the Nickel Plate ride a corridor of oily ties and white ballast through a sea of corn stretching green to the edges of the sky. Here and there a neat brick farmhouse, red barn, a silo, a grain elevator juts up out of the midwestern plain, casting long shadows in the morning light. Corn pollen sweetens the morning air. Gray-bottomed clouds billow on the horizon, and the distant muttering of thunder promise showers and a steamy afternoon.

A far-away wailing breaks the morning quiet, startling lines of red-winged blackbirds from their perch along the telegraph wires. The rustle of a thousand wings and piercing calls fill the air. A locomotive comes into view, gray smoke streaming from its stack and, unwinding behind, a string of coal-black hopper cars stretches back over the horizon.

The appearance of the giant instantly changes the quiet land. Birds and wind, even the sun, veiled in smoke, are subdued by its presence. The locomotive, wrapped in a cloud of vapor, exhaust chuffing and barking, siderods clanking, sends ripples out through the tassled corn. It quickly overtakes the two figures walking beside the tracks.

The two turn and squint into the whirlwind of steam and sound flying on spinning wheels. Each presses his cap to his head, the wind whipping their clothing. The locomotive rumbles by, shaking the ground, and the two wave to the engineer looking back from high up in his cab win-

dow. Steel rolling on steel, *clackclack . . . clackclack . . . clackclack*, oily steam and pungent creosote fill the morning air. "Fifty-five cars," the boy shouts above the clamor. Suddenly, the quiet morning returns as the caboose with its two flickering red marker lights flashes by and recedes into the distance.

"Number 600," the older man announces. "Worked on that Mike last year. Did you see how fast it was goin, Ben? It's a big, heavy one, all right. More firebox, 'n boiler 'n steam than you've ever seen. Got a booster engine on the trailer truck. By golly, that's one of the pullinest hogs ever." Alongside the boy, lean as a young tree, the older man's big shoulders, thick arms and gnarled hands express years of heavy work. That they are father and son is clear in the arch of their noses and the earthy brown of their eyes. They even walk alike, despite the years between them, and in the son's shoulders are already the lines of his father's stoop.

Meeting up with father and son this morning anyone'd know there was something special about the day. The boy's stiff, blue overalls, freshly starched and ironed workshirt, red bandanna peeking out of his back pocket and new black dinner pail swinging at his side mark this a day for beginnings, the start of an important summer, of a new job, of a new life. Today, at eighteen, Ben becomes a working man.

"It sure must make you feel good to see those two engines, a-knowin you worked on them. It's like . . . like they were yours." "Uh-huh, it's what makes the hard work good, Ben. You go along workin, not thinkin about it, not givin it a care. Then, one day, you see a locomotive you worked on, out here on the high iron, a-workin hard. Ya know, it still takes me by surprise. Even now, after all these years, I looked at that Mike a-steamin by and I thought, I made that one. Yep, I had a hand there."

Ben hears his father, but his gaze and his thoughts are already on the locomotive works just coming into view at the end of the tracks. "The Loco," they call it—a sprawling city of low, brown brick buildings with copper flashing weathered green. It sits in a triangle where the tracks of the Nickel Plate and the Baltimore & Ohio meet. Ten thousand panes of glass glitter gold in the low morning sun. In the corner of the big triangle, where the tracks cross Main Street, stands the seven-story main office, "one of the tallest buildings outside of Columbus," local folks boast. And beyond that, a gracefully tapered brick smokestack rises above the powerhouse.

The walk along the railroad tracks this morning has taken father and son from one world to another. They left their white clapboard farmhouse surrounded by corn fields, apple trees, and gardens of carrots and chrysanthemums for a new, hard-edged industrial landscape of brick and glass and steel. Another life begins here, where the high iron crosses Main Street. Here Ben's barefoot grasses and black earth become a separate landscape covered by rust-stained rock. The living green carpet that stretches in all directions as far as the eye can see ends abruptly at the steel and concrete forest. Signal bridges, storage tanks, cranes, semaphore signals, switch stands, coal silos, telegraph poles and signposts stand instead of trees. Each day, now, Ben will move back and forth between the two worlds, self-consciously at first, but soon as casually as the hundreds of other blue-clad workers converging on the main gate this morning in a murmur of accents.

Ben is no stranger to The Locomotive. He's come here more times than he can remember, beginning back when

he was a little kid. His dad brought him to the plant on Saturdays to show him where he spent his day and the wondrous things he was working on. It was during one of these visits he decided that someday, like his father and his Grampa Joe, he'd work here too.

◇

Ben grew up to the ticking of the old clock on the mantle, the squeak of the oven door, the honking of geese, the clanking of the pump handle at the side of the kitchen sink. Later would come faint voices and music crackling over a new radio. But the loudest sounds he knew were the thunderstorms that rumbled over the hot countryside all July and August. The brightest color Ben had ever seen, without doubt, was the red wagon with "ICE" lettered in gleaming gold leaf that stopped at the house often during the summer. The ice shavings that the iceman gave Ben to suck on were the coldest thing he knew, though an Ohio winter without your mittens was close.

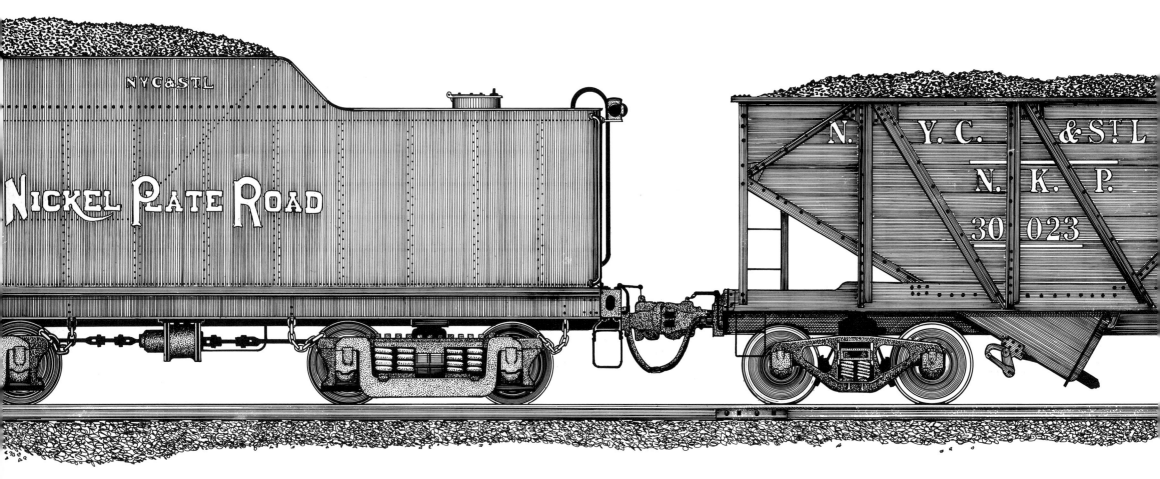

For Ben, life followed the rhythms of planting and harvesting. School let out just in time for Ben to help his father with the work of planting and cultivating. School began again after the corn and hay were in.

For Ben the arrival of fall had nothing to do with the position of the planets. First, he'd notice the nights getting cooler and the tinge of yellow and red in the maples. It would be darker when he got up to do his chores—feeding the chickens and milk cow—before leaving for school. And then, one morning, he would wake to the barking of all the dogs in Allen County (or so it would seem). In the distance, coming slowly his way, Ben would just be able to make out a familiar *chuff chuff chuff*, the rumble of wagon wheels and the sharp, piercing shrieks of a steam whistle. In an instant he would be at the window, and though all he would be able to see was a cloud of dust from which shot up puffs of black smoke, Ben would know what it was. There would be no school that day. For in a few minutes—Ben struggling with his overalls and high boots to get dressed in time—the puffing, hissing, roiling

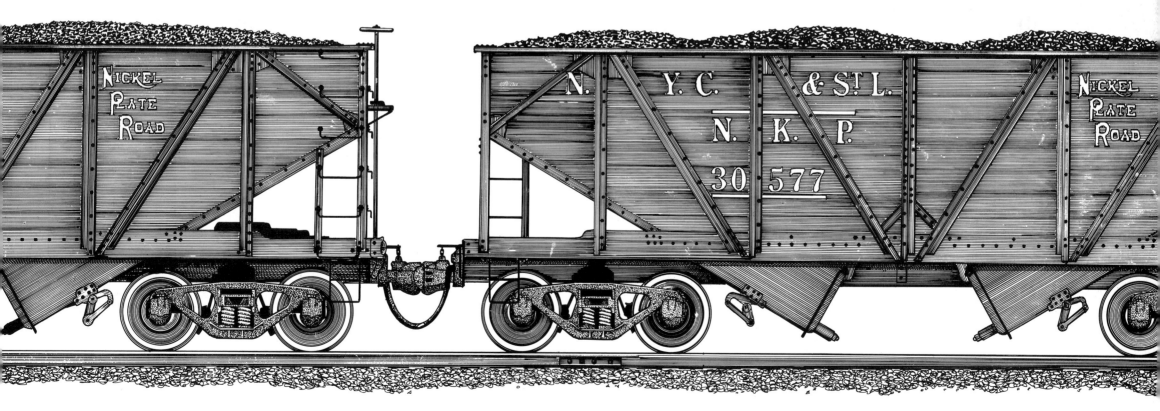

cloud would be at their gate. A black and green Case steam traction engine would lumber into his yard on six-foot-high wheels. Behind the engine, in tow, would come the threshing machine and farm wagons carrying a noisy, laughing crowd of men and women and kids. The custom thresherman would have arrived bringing many of Ben's neighbors to thresh the wheat piled in yellow mounds in the stubble-studded fields.

There was another cycle in Ben's life, less regular but just as certain. No sooner, it seemed to him, would he get used to having his father home on the farm day after day, than another cloud of dust would appear coming down their road. This time it would be an impressive black Buick bumping and bucking over ruts and mud holes, horn honking, rattling and sputtering right up to the house. Not many folks around had automobiles, so Ben would easily guess who it was. This time, as always, it is Mr. Larsen from The Locomotive.

"Hop in, Ben, and we'll go find your dad." Ben was happy anytime he had a chance to ride in an automobile (Ben always said "autoMObile"). He stepped high up on the running board and slid into the black cambric seat next to Mr. Larsen. The car quivered and trembled in time to its coughing engine. Ben pointed out his father and the two horses way out near a solitary oak. The Buick lurched backwards. And then with a chattering of gears it took off, right across the field. "How fast can it go, Mr. Larsen?" Ben always asked the same question, and was always impressed with the answer. "Forty. It'll get up to forty on a straight, smooth road." Mr. Larsen's voice wavered and he swallowed a couple of his words as the automobile jolted over the furrows.

Ben's dad called the team to a rest in the shade of a spreading oak, dropped the reins and climbed down off the spoke-wheeled plow and walked over to meet them. They talked about the weather and crops before Mr. Larsen

brought up the subject of the new locomotive order. Louis looked down at the fresh, black furrow, kicked at a clod of earth and scratched his head the way you do when you're thinking. "Niels," he said suddenly, with a smile, "tell me how it is that the only time the New York Central needs a batch of engines is when I'm just about t' get my corn into the ground." They laughed and fell silent, and all Ben could could hear now was the restless stomping and whinnying of the big Percherons. "A new locomotive, you say." Ben's dad looked down again and kicked the toe of his boot into the rich, dark soil. "Somethin' new, huh?" He thought another moment, took a deep breath and let it out. "Well, Niels, give me a few days t' get my corn planted and I'll be there. You know, I wouldn't miss it for anythin." He turned to Ben. "Well, son. We're goin to build us another engine." Ben was excited. It would mean visits to The Locomotive again. But he was sad, too. Somewhere inside he suffered a deep pool of loneliness whenever his father went away to work. "But I'll miss you dad," he thought, though he never said it.

There was never any doubt in Ben's mind that someday he'd go to work with his dad. He grew up with locomotives, powerful, immense creatures that always scared and fascinated him. He had lived all his eighteen years within hearing of the tracks, falling asleep each night with the distant, lonesome whistle of a drag freight.

When Ben was little he would lie in bed in the morning, the soft, frayed blue-and-white bearpaw quilt pulled up around his shoulders, listening for the 6:10 to head out of the yard. He could hear the bump of the locomotive coupling up, and the engineer pulling the slack down the long line of cars. Then would come the rapid, staccato bursts of exhaust as the drivers slipped on the frosty rails. "Needs more sand," the boy would say to the imaginary fireman in his dark quilt cab and, sure enough, in a moment would come the slow, plodding *puff . . . . . puff . . puff puff puff*

puff puffpuffpuff as the drivers took hold and the heavy freight began to roll. Ben was in that cab, his hand resting on the throttle, his eyes searching the mist for signals, his fireman shoveling rhythmically at his side, the cold morning air stinging his face, as his locomotive slipped through a confusion of switches and headed out to the high iron.

Trains filled Ben's life and were the stuff of his fantasies and play. Uncle John Keller used to take him to the yard with his boys, and once they rode a caboose the twenty miles all the way to Delphos and back. He put them up into the cupola where the brakeman sat so they could look along the tops of all the freight cars the full length of the train. Ben especially liked the flatcars loaded with shiny new plows, disks and mowers lined up in a row.

As Ben sat there on the worn leather seat box looking out over "his" train, he imagined a driving storm, the right of way and black fields of corn stubble blanketed with snow, and him, grown up, seated inside the caboose warmed by the potbelly stove. The tight bunks promised him a warm night's sleep. The caboose smelled of wet wool coats, kerosene, pipe tobacco, coffee and coal smoke.

The train stopped. It was so quiet now Ben could hear the snow tick against the windows. He waited for the one long and three short blasts of the locomotive whistle that would send him out into the fierce storm with his lantern and fusees to protect the rear of the train. He imagined the icy cold of the plains winter, snow blowing in his face. He'd stamp the cold out of his feet and clap his gloved hands together, peering out into the dark night beyond the circle of red light sputtering from his fusee. Like the brave brakemen he had read about, he would protect the rear of his train no matter what.

Ben was sure he was about to freeze to the bone, when four long blasts of the whistle called him back to the warmth of the caboose. Now he imagined how good it felt to return to the yellow glow of the kerosene lamps and have a hot cup of coffee from the smoke blackened blue-and-white speckled pot on the stove (though he hated coffee, even with mostly milk).

Ben's imagination lived on rich images gleaned from the stories of railroad men he knew and the pages of books and magazines. "Drink Coca-Cola," read the advertisement he had pinned up on his bedroom wall, "and learn why railroad men find in it the acme of wholesome, delicious refreshment." There, below a bottle and glass of Coca-Cola, sped a high-wheeled locomotive pulling a crack passenger train, and in the cab a smiling engineer waving to Ben, waving to every boy who wanted to be an engineer when he grew up.

Ben, and many of the boys he knew, lived trains. Even in school, no matter what the lesson might be in his history class, Ben always had his textbook open to a photograph of the *Empire State Express* which looked about to roar right off the page and into the classroom. He liked waiting for a haircut at Mr. Kruger's barbershop where he would eagerly dig through the old copies of *Railroad Man's Magazine*, *Railroad Stories* and *McClure's* to read and re-read tales about railroading by his favorite author, Frank Hamilton Spearman.

When it came his turn, Ben would climb up into the big chair and onto the padded board Mike put across the arms for him. Then, with a flourish, Mike would throw the pin-striped cloth around in front of Ben, and pin it together above his collar. All the time he was tucking the little tissue paper collar around Ben's neck and snipping at his hair, Mike would tell him the latest stories he'd heard from railroad crewmen, all in his precise German accent.

"Dis engineer told me yesterday, Ben, dat he and his fireman vere caught Tuesday night in the vorse storm ever. A passenger train it vas. And it vas raining so hard, the headlight could not see through the vater pouring down. Pitch black it vas." Mike snipped carefully at Ben's sideburns and put his mouth close to the boy's ear. "Den it vas dat the fireman called out to the engineer dat der vas supposed to be a signal back a vays and he hadn't seen it." Ben's hair bristled and goose bumps popped out all over his body. "Vell, I tell you, dat engineer stopped that train right now, he did, and sparks flew from those big drivers like fire." Mike interrupted his story to wave to the postman passing by outside the big window. Ben squirmed impatiently. "He vistled for the brakeman to protect the rear of the train. *Tooooot. Toot. Toot. Toot.* Den he signaled the head brakeman to go out and protect the front of the train. *Toot. Toot. Toot. Tooooot.* Vell, dat head brakeman vas gone only a minute and den he vas back lickety-split, and he tried to call up to the cab but no vords vould come out of his mouth." Mike interrupted his story again, this time taking too long, Ben thought, to choose just the right clippers layed out next to the straight razor and shaving mug on the clean white towel. "Vell, the engineer and fireman got down and valked to the front of the train and saw vat the brakeman saw. Not ten yards ahead, Ben, not any farther than from here to dat coal wagon out there"— Mike's voice turned into a hissing whisper—"the tracks ended. Ya, just like dat." The barber's hand sliced down through the air like a cleaver. Ben's heart pounded like it would leap from his chest. "The bridge vas gone!" Ben gasped. "Ya, und a hundred feet below, the raging, roaring river." Mike opened the safety pin behind Ben's neck, took off the little tissue paper collar and shook the cloth away, showering locks of black hair onto the floor. "Zo, a good haircut. Your mamma and pappa will like it." Ben was stunned, the last words of the story ringing in his ears, as Mike brushed off his neck in a cloud of perfumed talcum powder and helped him down to the floor.

At Ben's grandparents' there was a special pile of tattered books set aside especially for him to read when he visited the big gabled house on Catalpa Street. Allen Chapman's *Ralph of the Roundhouse*, dogeared and turning yellow at the edges of the pages, was one of his dad's favorites. Now Ben would spend hours in the itchy overstuffed green chair with white-lace antimacassars pinned to the threadbare arms, savoring the stories and pictures, his grandmother's marmalade cat warm and purring in his lap.

One day, a cold, darkening winter afternoon that began his Christmas vacation, Ben was passing a brightly lit shop window on Main Street when something wonderful caught his eye. There was the first electric train he had ever seen, a whole Lionel layout. He dropped to his knees in the snow, pressed his nose to the frigid glass and watched the little green electric locomotive with gleaming brass window frames speed by only inches away, headlight glaring ahead.

For Ben, time stood still. He had no sense of the cold and dampness penetrating his wool knickers. Crossing gates moved up and down and tiny red lights flashed back and forth at grade crossings. The locomotive started and stopped at miniature semaphore signals. Everything was enameled in bright, glassy colors. Pressing an ear to the window he could hear the faint clatter of the wheels on the tracks as the little trains sped through lift bridges and tunnels in tinplate mountains, passing bungalows, passenger and freight depots, interlocking towers and a powerhouse topped with a tall brick stack. Ben made all the chuffing noises needed to bring the scene alive.

The little trains seemed utterly real to Ben. A coal car dumped its miniature load alongside the tracks, and on a side track waited a wrecking derrick with a boom that really moved up and down. The passenger cars, caboose, the signals, the depot and powerhouse were all illuminated and everything could be made to happen by "Distant Control," with just the push of a button. Ben returned to that window often until one day, after the New Year, it was gone.

Caught up as he was all his life in the romance of the rails, it's a wonder that Ben didn't start working at The Loco long before today. Actually, he had tried, the summer before his junior year in high school. Ben and his buddies, Steph and Christian, had been going to the plant often to visit fathers and grandfathers, uncles and brothers and school friends who worked there. They'd watch new locomotives take shape on the erecting floor and—this was their favorite event—watch the overhead crane pick up an entire locomotive, one of the big Mallets—boiler, wheels and all—and carry it, siren howling, from one end of the erecting floor to the other. "How much does that locomotive weigh?" Ben once asked the man directing the crane. "Three hundred tons," he answered, "but that crane will hoist twice as much."

The more time Ben spent at the plant the harder it became for him to stay in school. He thought more and more about being on his own and having a job, being there in the shop with his father and grandfather and his friends, and being a machinist or maybe even becoming a draftsman and working in the engineering department. He was sixteen, which is the age most boys started at The Loco. (Years ago, when his dad was growing up, kids started at eleven or twelve.) It seemed to him he had had enough schooling. After all, he was better than most with numbers and as good as a boy needed to be with writing and spelling and such. Mr. Baller, his mechanical drawing teacher, was already giving him senior drawings to do, intersections and developments, and hard assignments like that. Besides, most boys his age weren't in school anymore, and it just seemed like a good time to have a talk with his dad about a real job.

So Ben mentioned it at supper one night. "Y'know, dad, a lot of the guys at school are going to work now—Steph, n' Harry n' Nikos." His father suddenly became more serious than usual. And his mother suddenly remembered something she had to do right at that moment in the kitchen. All of a sudden Ben seemed so grown up. Until this moment he had been a boy in her eyes. A boy who only yesterday couldn't get his schoolwork done on time, who had to be reminded half a dozen times to do his chores, who tripped over the dog and broke something every time he went into the kitchen. This boy was going to work and, God help us, would be put in charge of running a big, expensive, complicated machine.

"And . . . well, I think I'd like to go to work too. So, I thought you might ask the foreman about it, soon, maybe tomorrow." Ben began laying out all the reasons he had carefully rehearsed in his head for days, talking as grownup as he could and trying not to be deterred by the look on his father's face. Yes, his dad agreed, Ben was strong enough and responsible enough to have a job. And he was certainly good with numbers, and as good with words as a boy needs to be. And yes, he had just gotten an award for lettering in his mechanical drawing class. And hadn't his dad complimented him many times on the work he did on the small lathe and drill press in their basement workshop? Ben pointed out several times that the job would help with the family's bills and make it unnecessary for him to get an allowance. "You're a-growin up Ben. Can't deny that. And you'd make a good machinist, all right." Ben could hear the hesitation in his father's voice. "But, your mother and I feel your place is in school right now. Yep, that's where you ought to be right now. You'll get to work in good time, son, in good time." And Ben knew the subject was closed.

Ben was the first in his family to get this close to graduating from high school. What's more, he was doing well. His Grampa Joe hadn't had any schooling at all in the old country. As the old man delighted in telling, he had grown up in a blacksmith's shop in a small village in Finland, where his father—Ben's great-grandfather—had made and repaired farm implements and all sorts of things of iron.

For many reasons Ben's dad thought it important that his son complete high school. Ben loved Grampa Joe and his old-world wisdom dearly. He respected his skill and realized that he probably couldn't find in himself the old man's patience. And he was awed by the way this simple man moved in and out of shops in Lima speaking Russian, German, Polish, Finnish and fine English. But he sensed, also, that his dad wanted something more for his son, more than he and Grampa Joe had. This was America in the 1920s, not Grampa Joe's village Finland of another century. And somewhere in the crossing, educated sons had become important to fathers. These days, what was important to your dad becomes important to you too.

Still, Ben was disappointed, even more so when he met Steph at school the next day. Steph's dad, it turned out, had agreed to let him go to work and promised that he'd talk to his pool leader about it. The boys felt sad at the turn of events. Ben conceded that there was no way to change his father's mind (and, though he didn't say it to his friend, he understood and really didn't want to). Nor could they even imagine Steph giving up a chance at the job they both dreamed of. Somehow, the more they talked about it the brighter things seemed. There was still time to do the things they always did together. And Ben suspected that not even a job would make Steph less mischievous and funny. Besides, they had been friends so long—since first grade—it wasn't likely anything would change that now.

Ben and Steph looked at each other and started to laugh and horse around like they usually did. With that gleam in their eyes that meant both of them were thinking the same thing, they headed off to Mr. Wills' drugstore. There, they climbed up on the tall revolving stools in front of the green and black tile soda fountain and ordered chocolate sundaes with extra Hershey's syrup, vowing that good fishing and baseball buddies would never part. "It's just a year, Ben; just a few months, really," Steph assured him, "it'll go by before you know it."

◊ *2* ◊

Apprentice Kid

BEN FINDS his time card in the varnished oak rack on the wall along with his dad's and his grandfather's. There is his name, lettered in ink and an accountant's bold hand at the top of the buff card, and his rate of pay—25 cents an hour. The card feels good in his hand, heavier than it really is. He looks at his name a moment longer. There's something about it that seems more important this morning. Having watched the other men in line in front of him, Ben knows to slip his card into the slot under the big clock face and pull down the handle. Then he drops the card with his first day's starting time casually into the "In" rack as though he has been coming to work here every morning for twenty years.

The appearance of the time card means a lot to Ben. It is a sign that he was expected, that he really has a job here, belongs to The Loco and this group of men he so admires. When the boy turns to await his father coming through the line he notices that his dad and the other men close by have been watching him. All of a sudden Ben realizes that he's wearing his boyish grin, the one that just pops out at moments like this and, thinking it won't do for now, draws his body up into a more dignified pose. He sees reflected in his father's face the glow of his own pride this morning. "It's sure good to have you here, Ben," he says, quietly enough so no one can overhear, and reaches up to put his arm around the boy's shoulders. Ben sees tears glisten in the corners of his father's eyes.

On their way to the locker room, Ben is still thinking about his time card when his dad takes him by the arm and leads him into a small, cluttered office. Every inch of floor is covered with piles of catalogs, stacks of specifications, small castings, patterns, tools and machine parts. Every inch of the walls is covered as well with several layers of blueprints, pencil sketches and locomotive portraits in heavy mahogany frames. There must be a dozen calendars, advertising Pond Machine Tools and Starrett Gauges, and *The Twentieth Century Limited*, coming through a cloud of steam right at you. Here and there are little deckle-edged snapshots of kids with toothless smiles, of company picnics, Dalmatian pups, a fisherman proudly showing off his string of trout. An iridescent blue ribbon proclaims the winner of the county fair fly-casting contest. Against one wall is a huge roll-top desk stacked with blueprints and letters and notes weighted down with gears, twist drills and bolts. Hidden behind the busy clutter, revealed at first only by the smoke rising from his pipe, sits a lean, spectacled man even taller than Ben feels today.

"Ben, I'd like you to meet the plant foreman, Mr. Schnell." Ben shakes the foreman's rough, workingman's hand firmly as his dad has taught him. And when he looks straight into Mr. Schnell's blue eyes he's surprised by a boyish gleam. Ben likes him already. "Pleased to meet you, Mr. Schnell."

"You've got a good, manly handshake there, young fella.

It's always good to meet the son of a man like your dad. He's the best machine shop foreman I've ever had the pleasure to work with." Mr. Schnell leans over closer to Ben's ear and lowers his voice to a loud whisper. "He won't ever let on, of course, but *I'll* tell you."

Mr. Schnell pulls open a bottom drawer in the big desk, rummages around a minute and comes up with a sheaf of time-yellowed papers. He blows off half a century of dust and shows Ben five pages bound together with a loop of string. It's Charlie Schnell's indenture papers. He was sixteen, and it was an Ohio June just like this one, in 1885. Now he looks over the packet fondly, like he's found a long-lost friend, touching a finger to his tongue before turning over each page. He hesitates, and then looks at Ben. For a moment he sees himself, young Charlie, painstakingly signing the indenture in his schoolboy's script.

"With that signature and my dad's, Ben, I bound myself for three years." Mr. Schnell searches for a particular paragraph and then reads it to Ben. "Yes sir, this young man agreed to 'behave himself as a faithful apprentice ought to do.' And that's spelled out here too. 'Not be absent from services of the said company during all or any of the proper working hours without leave first obtained . . . keep himself free from all bad habits and wholly abstain from the use of vulgar, obscene or profane language, and shall not use intoxicating liquor as a beverage.'" The foreman held out the indenture papers for Ben to see, and continued reading. "In exchange, the company promised to 'employ good and competent instructors to give said apprentice proper instructions in the use of tools,' and to teach him 'the trade and skill of a machinist.' And see here, my pay began at six cents an hour." Ben read the part that said that if Charlie Schnell successfully completed his apprenticeship he would receive the certificate of a Journeyman Machinist, a kit of tools worth fifty dollars and 'the privilege of remaining in the employ of the Company.' "And look here Ben. Here's my father's signature. My, my, how long it's been since I've looked at this."

Charlie has stayed on for forty years. And during those years he's worked at almost every job there is to locomotive building. The men refer to him as a "mechanic's mechanic," a master to whom the most experienced man in the shop can come with a problem that stumps him. Ben's dad told him that Mr. Schnell doesn't always know the answer, but he knows how to work it out, and how to get you to come up with the answer. "Mr. Schnell made you realize that you actually knew the answer all along. What he really did was help you find it inside yourself."

The foreman is interested in the young helper, and asks him about his experience. He finds out the boy already has a working knowledge of the lathe and drill press, that he knows how steam locomotives work. Charlie Schnell is sizing him up, making the questions harder, finding out just what Ben knows. To Ben it seems less like a sizing up than just talking about things that interest him, the way he always does with his dad and his grandfather at the supper table.

Then Mr. Schnell stops asking questions and just looks at the boy for a moment, with a critical eye, the way he looks at a new design just come down from the engineering department. "Young man, how would you like to work for me as a helper?" Ben looks to his dad who knows that the question is only a courtesy. When Charlie Schnell asks how you'd like to do something he is really saying that he wants you to do it. In his surprise the boy can only smile and eagerly nod. "Yes . . . yes, sir, I'd like that very much, Mr. Schnell."

"Well, that's fine, Ben, just fine," the foreman announces as though an important deal has been struck. "You start right now. I'm going to have you mount this set of blueprints on boards and then take them up to the machine shop. But before you go I want to show you what they're all about."

Mr. Schnell regards the ten-foot-long blueprint on the wall with reverence, approaching it the way people do a valuable painting in a museum. "What's the scale of this drawing," the foreman asks. Ben looks immediately to the title block in the upper right-hand corner: "One-and-a-half inches to the foot, sir."

"And what does that mean," Mr. Schnell asks, continuing his good-natured but serious test of the new helper's knowledge.

"Well, sir, it means that every inch and a half on the drawing equals one foot of real locomotive . . . and that this is one BIG locomotive."

"And maybe the most efficient and powerful that Lima or anyone else has ever built." The foreman taps the blueprint with the amber stem of his pipe. "Look at this firebox. A hundred square feet of grate. Why, that's half again the area of the grate in the big Mikes. And see here? The firebox is so big that they've come up with a new trailing truck to support it, a two-axle truck."

"Which makes it a 2–8–4," Ben and his dad say almost in unison. Ben feels a sudden tingle. Through his mind flashes images of the Mike they saw on their way to work this morning. How new, how modern it seemed, and how proud his dad was of it. Now, all of a sudden, something had changed. This new locomotive, Ben could see in the crisp lines of the blueprints, makes the 600, his dad's Mike, seem old-fashioned. And it had all happened in the short space of a morning.

"Those are William E. Woodard's initials on the print and this promises to be some locomotive," Mr. Schnell says pointing out a couple of interesting parts to Ben and his dad. "So you keep your ears and eyes open out there in the shop and you'll find out a lot about Order No. 1070 over the next few weeks. For now, it's time to get these prints mounted and then up to Joe Park in the machine shop loft. And while you're there, spend some time with Joe. He's got a lot to teach you."

Being a "helper" is everyone's first job at The Locomotive. That's how Ben's grandfather and his dad started out and Mr. Schnell, too. And it's significant to Ben. Even the plant foreman, "the mechanic's mechanic," has done the very things Ben is doing today. In fact, his dad has told him, some of the engineers who design locomotives and whose drawings Ben so admires and tries to emulate started out just the way he is.

A helper's days are spent running errands and giving a helping hand wherever it is needed. He gets a lot of exercise. The Locomotive is forty-nine buildings spread out over sixty-four acres. He might be sent from the machine shop to the erecting shop with a part, pick up blueprints at the main office and deliver some to the boiler shop and some to the hammer shop. He could start out on a errand for Mr. Schnell and never really know how it all might end. But that doesn't matter. Everything is new and there is so much to see and learn.

At the end of his first day Ben's head is spinning with all the new things he's seen and heard. He talks through supper and in his excitement forgets to eat until his mother reminds him that his food is getting cold on his plate. Ben's dad recalls his first day of work for his son, and a lesson he learned that day. "I guess I was about thirteen. One of the guys called to me from across the shop. He was holdin up a bright red piece of fruit and gesturin to me, wantin to know if I'd like it. Well, I thought, that's mighty friendly. Here it is my first day and all, and I don't know this man and he don't know me, and he's offering me a nice red apple. So I hold up my hands, and he throws it from across the room and when I catch it —SPLLAAATTTT. It weren't an apple at all, but a big, red, ripe mushy tomato." Ben burst out laughing like he

would split his sides. "Yeh, well, all the other guys got a good laugh, too, when they saw your dad standin there with a big surprised grin and tomato all over his face." Ben could tell that the memory of it was still so vivid for his dad after so many years.

"And you know who that guy was, Ben?" Ben shakes his head. "It was none other than Charlie Schnell, that's who." Even Ben's mom laughs, and Ben is pleased to see her smiling again. He doesn't understand why she has been so quiet and wistful through such a happy evening.

◇

Ben would never think of himself as a historian. That wasn't exactly his favorite subject in school. In fact, he had spent most of his time there daydreaming to the distant chuffing and shunting and whistling of locomotives that reached him through the tall windows open wide for warm spring and fall days. (He had always volunteered to be in charge of the long window pole with the hook on the end just so he could be sure the windows would be open for him each day.) But that night, at home, in the front porch swing, he thinks about change and growing up, about following in the footsteps of his father and grandfather, about a lot of things historians might think about.

Mostly, Ben is thinking about that morning, about walking to his first day at work with his dad, about the locomotive they saw. His dad was so proud of the big, new Mike. Ben had caught it in his voice. And there was the way his dad seemed to reach out and touch every part—the gleaming brass bell and whistle, the white-rimmed drivers, the burnished steel rods and valve motion—with his eyes. He knew them, loved them in a way like . . . like one of his children.

The dusty sky darkens to a raspy chorus of frogs from

the duck pond. Ben is thinking more and more about the new locomotive. What is it about that? It isn't even a locomotive yet, just an idea—white lines on blue, that's all. But it intrigues him. Somewhere, in the darkness beyond the screened porch, horse's hooves clatter on paving bricks, cicadas buzz and click, an automobile toots. Lilacs perfume the muggy midwestern evening. A familiar far-away whistle announces the late hour and time for bed.

Alone in the dark, Ben the historian is gradually beginning to grasp the meaning of No. 1070. This had been his first day at work. And hadn't Mr. Schnell asked him, Ben, the apprentice kid, to mount the first set of blueprints for No. 1070 and take them to the machine shop? Much of his first day at The Loco had to do with the new Mike with the extra axle in its trailing truck and a powerful new look. Well, there it is. The two of them, Ben and No. 1070, were starting out together. Just as the Mike was his dad's, W. E. Woodard's new locomotive is his.

⋄ 3 ⋄

What Next, Will Woodard?

THE STORY of Ben's locomotive begins in the imagination of a slight, energetic man who has spent much of his life thinking about steam power. W. E. "Will" Woodard learned locomotive design from the best of an older generation of engineers. Now, at the Lima Locomotive Works, he is refining what he has learned to create new locomotives, more powerful and more efficient than any built before. To find out how he does it, we'll need to leave the shops and spend some time at the engineering department up on the fifth floor of the main office building.

The engineering department is where ideas like power and efficiency take shape in equations and drawings. Row upon row of basswood drawing tables tilted up every which way fill the sunlit room. At each table a draftsman works up an engineer's sketches and notes into detail drawings for the machinists, pattern makers, forgemen and boilermakers. The black outlines of graceful, spoked drivers, rounded domes, flared bells and a thousand other parts flow from compass and ruling pen nibs onto light blue tracing cloth. One draftsman, twirling a drop compass between his fingers, inks in the hundreds of staybolt holes to be drilled in a firebox sheet. Another, leaning on his wooden T-square, his eyes close to the drawing, inches a yellowed celluloid triangle across the board to create the fine threads of a machine screw. A red-haired boy, perhaps all of twelve, moves up and down the rows delivering rolls

of blueprints, notes from the shop and fresh pieces of vellum and tracing cloth to the men working at their boards. The tall north windows look out over a busy railroad scene. Steaming by are locomotives which only a few months ago were drawings on these boards.

Somewhere among them, Will Woodard is bent over a drawing, head to head with a young draftsman. His ideas flow out through his hands into delicate pencil sketches, dimensions, quick calculations. His manner is quiet, retiring. "Now, you want to put plenty of metal here and here," he cautions, pointing out stress points in a massive steel casting. "Large radius fillets in these corners will give us the section and strength we want." Will looks over the draftsman's careful linework. He picks up a scale and checks several dimensions. "You know, Tom, figuring the draft for this piece is going to be tricky." Will taps the drawing with his finger as he thinks. "Why don't you take your sketches down to the foundry and get some suggestions. Marko has a feeling for these things. We could work half the day trying to make sense of it all from these tables, but he just knows, he has it in his hands."

Will leaves the draftsman with a touch on the shoulder and returns to his own board. Adjusting the silver-rimmed spectacles on his nose, he pages through a pile of fresh, clean blueprints until he finds what he is looking for. Slowly, as he unrolls the drawing, a new locomotive appears from pilot to cab. The title block says simply "Elevation No. 1070," but against the flag blue background is the white outline of an extraordinary locomotive. Will touches the print here and there, feeling the cool, gray steel.

◇

Ben has heard a lot about this Mr. Woodard. He had met the engineer through his drawings that very first morning in Mr. Schnell's office. He's always wanted to meet the man who designed "his" locomotive. But Mr. Woodard isn't around much. He spends most of his time at the Company's headquarters in New York. It just seems to Ben that he and the famous engineer who works up in the big office building inhabit two different worlds. Ben is, after all, only a helper. His days are spent sweeping around the machines, hauling metal shavings, running errands.

Still, hardly a day goes by without some new adventure. And one day, when he is running some corrected prints to Mr. Schnell's office, it happens. The foreman looks over the prints, puts his initals in a box up in the title block and thrusts them into Ben's arms. "To the engineering department," he says turning back to his work on the big cluttered desk. But Ben remains, hesitating, uncertain. Mr. Schnell looks up at the uncomfortable lad still standing next to the desk. "Well, what's the problem young man?" Ben explains, haltingly at first, that he has never been to the main office. And, well, is it all right to go there, like this, in greasy overalls? Mr. Schnell smiles. Ben, so tall and strong, was still very much a little boy inside. "You look fine Ben, just fine. Engineers are no strangers to shopmen in greasy overalls. Besides, I suspect many of them would rather be out here in the shops, with the tools and the locomotives."

When he arrives at the engineering department, Ben is still feeling a bit uncomfortable, but his curiosity takes over. He stands at the door a moment, takes a deep breath and steps in. Draftsmen work quietly at their boards. A group of engineers talk over a long table covered with huge drawings. Ben realizes he's unnoticed and begins to explore the busy room. A print unrolled on a nearby board catches his eye. It's an eight-foot-long erecting drawing for a tender. Up in the title block is the now familiar "No. 1070." It's the new twelve-wheel tender for Ben's locomotive.

The apprentice kid is enchanted by the collage of sketches, penciled notes, adding machine tapes, small prints, erasers, triangles, scales and overstuffed file folders covering the board. Ben jams his clinched hands down into his overall pockets to keep from touching and holding the satiny German silver compasses, dividers and ruling pens, some in a fine black leather case lined with shimmering green velvet. He moves around the board to get a closer look at what seems to be some complicated kind of scale. It is a beautiful thing with fine lines and numbers so precisely engraved into a finish like bone, and a little glass plate gleaming like a jewel. Ben is so absorbed that he doesn't even notice the small, balding man who comes up behind him and moves quietly to his side. Will Woodard climbs onto the tall stool and, seeing the boy's fascination with the instrument, places it into his hands.

"Have you ever used a slide rule? It is really quite simple, you know." William Edward Woodard (Cornell University, 1896) and Ben the apprentice kid (Lima Senior High, 1924) introduce themselves and, without another word, begin a lesson in high speed calculation. "Do you remember how to extract the square root of a number?" Ben couldn't remember exactly. What he could remember was standing up at the blackboard in Mrs. Foster's arithmetic class working on a column of numbers that began in a radical sign above his head and ended at the chalk rail. It seemed to take hours, especially under Mrs. Foster's piercing gaze which seemed to burn into the back of his head.

"Let's say we want to find the square root of 2. We put the hairline precisely on the 2 on the A scale. Then we simply read the square root off the D scale—1.41. Go ahead, try it. Find the sqaure root of . . . 10."

By the time Ben learns to read cube roots from the K scale and to multiply and divide on the C and D scales, he and Mr. Woodard have talked about Ben's interest in the new Mike with the four-wheel trailing truck and about his awards in mechanical drawing. And they've become friends. Ben takes Mr. Woodard back to the joys and wonderments of his own childhood. Mr. Woodard shows Ben, as though holding up a mirror, a world of possibilities.

Will Woodard rummages about in a cluttered desk drawer for a moment and comes up with a student's wooden slide rule, which he gives to Ben. "You've made a good start, so now you can practice what you've learned today. I used that in college, but it has been lost in the drawer since I got this new one. I'm sure it will enjoy being in use again. And, Ben, when you get back to the machine shop, tell Mr. Schnell I need you tomorrow for an hour or so."

◇

When Ben returns to the engineering department (in clean, ironed overalls, today) he finds Will Woodard at his drawing board. He isn't certain what to expect. Maybe, he thinks, Mr. Woodard has some errands or some other work a helper might do. Ben is excited about the prospect of just working around this man he's heard so much about. But, as it turns out, the engineer has something special planned for him. "Ben, I think you need to know something about how we design locomotives, so you can better understand what's going on out in the shops." Ben can hardly believe his ears. "Here, climb up on this stool, and we'll just roll up our sleeves and get to work."

"There's a lot of history behind those big locomotives you're building out there. You know, when I was a boy, growing up in Utica, New York, locomotives were half the size they are now. The heaviest of them weighed maybe 150,000 pounds. Of course, I thought they were giants. But I look at what we're building today—Like that 2-8-4 of yours—and I can hardly believe it. That engine will weigh almost four hundred thousand pounds. And that's just the beginning. Not many years down the line you're going to be working on locomotives that weigh nearly a million pounds! Think about that, Ben, a 500 ton locomotive."

"I can't imagine a locomotive that big."

"Well, you just wait and see. The railroads need more motive power, Ben. There's a serious problem out there today. The freight business is booming. More and more people are riding the comfortable new Pullmans and coaches. Freight trains have reached their practical limits. Long freight drags, doubleheading with two locomotives

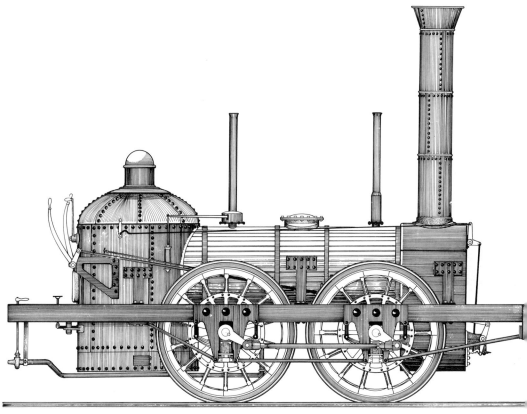

John Bull, 1831

up front and sometimes even a pusher behind, trudge up steep grades slowing traffic to a snail's pace. Traffic is bogging down. The railroads are pinning their hopes on faster, more powerful locomotives."

"Like the 2-8-4."

"Like the 2-8-4. And I'm already at work on a 2-10-4. But it's not simply a matter of building bigger, more powerful faster engines. We've got to think about weight."

"I was thinkin, Mr. Woodard, that a million-pound locomotive is goin to have to have some pretty hefty rails under it."

"And that's good thinking, Ben. We've always used a simple formula for more power. Beginning with the first little teakettles of the 1830s, builders increased the power of locomotives simply by making them bigger. A bigger firebox can hold more wood or coal and a bigger fire. A bigger fire can heat a bigger boiler full of water to steam faster. More steam and bigger cylinders means more power. Back when I was just starting out at the Baldwin Locomotive Works things were simple: need more power? Build a bigger engine."

"And the bigger they get the heavier they get, right Mr. Woodard?"

"They have to, Ben. Locomotives have to be heavy. Try to imagine that place where the round, smooth steel tire of a locomotive drive wheel rests on the polished rail head." Will hands Ben one of his compasses. "Here, draw a circle. That's a drive wheel. Now, with the T-square draw a line that just touches the bottom of the circle. That's the rail."

"The line is tangent to the circle." Will raises his eyebrows high over his glasses and looks over at Ben as though he's just discovered him sitting there. "I learned about circles and tangents in Mrs. Foster's class."

Consolidation, 1866

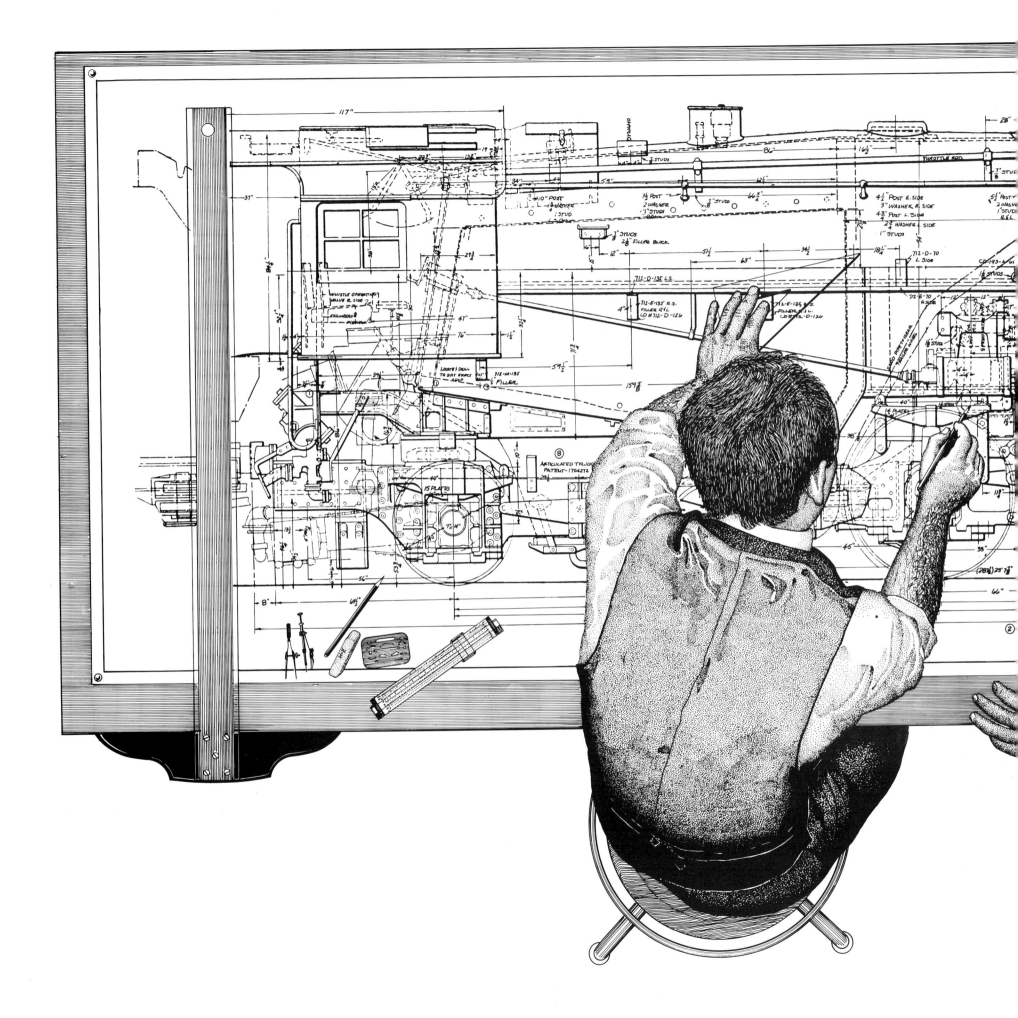

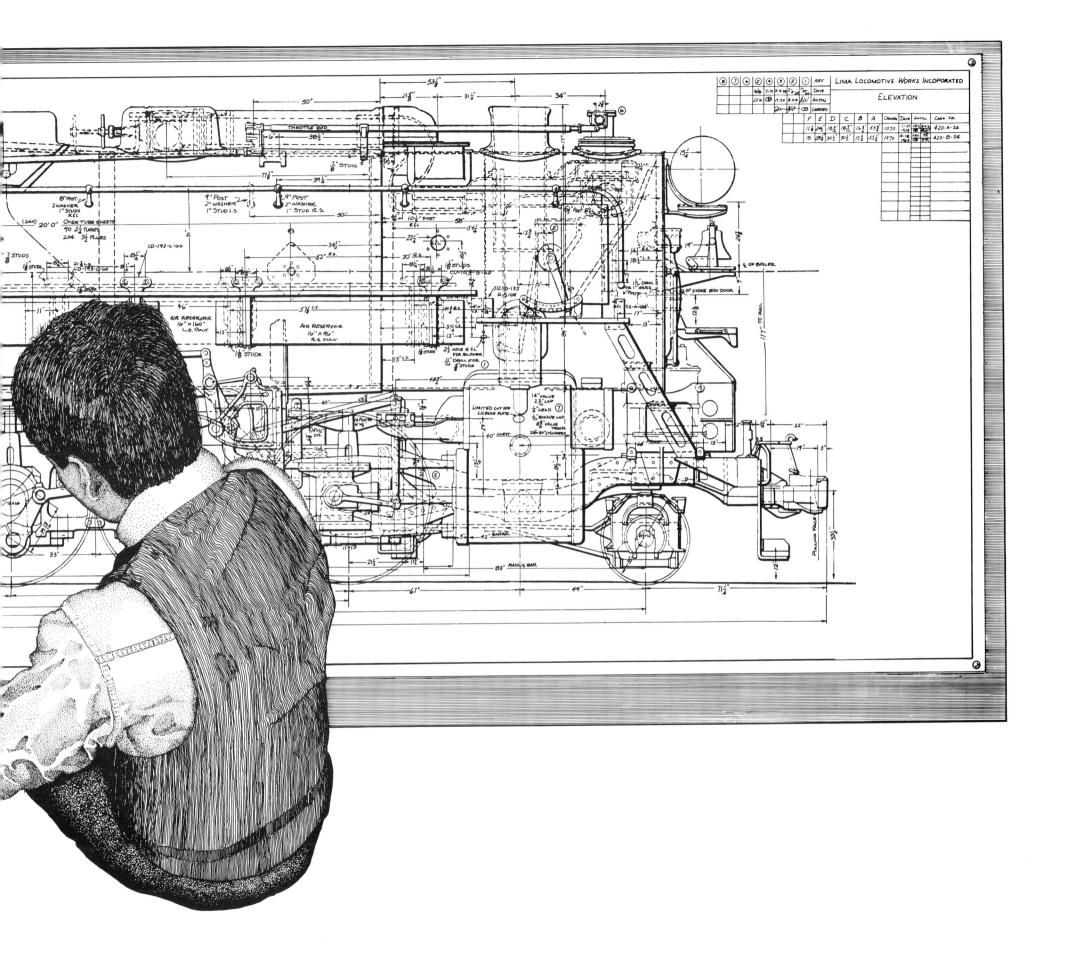

"You have indeed." Will looks even more intently at Ben. "You have indeed, Ben. Well, that point where the line is tangent to the circle isn't very big is it? Even if a locomotive has eight drivers, the total area where the tires touch the rails is only a few square inches. Engineers and enginemen know these few inches well. When the engineman opens the throttle and the drivers begin to turn, and there's a long, heavy string of freight cars behind the tender, everything happens right there in those few square inches.

"We have a term for describing how much the locomotive tires grab the rails when starting up: 'adhesion.' If the adhesion is too low, the drivers slip on the rails and the locomotive and train go nowhere. As adhesion increases, the drivers take hold and grip the rails and pull the train forward. Adhesion increases as more and more weight is put on the drive wheels."

"So to make a locomotive more powerful you make everything—firebox, boiler and cylinders—bigger, add another pair of drivers and put as much weight on them as you can."

"Up to a point, Ben. There's a practical limit to the size and weight of a locomotive, and we've reached it. As locomotives grow all the things around them have to grow too. Heavier rails, first of iron and then steel, were needed to support all that weight and stand up to the pounding of the drivers. Back in the 1830s and 1840s trains ran on light rails weighing 25 pounds per yard; today, out on the high iron, they're using 131-pound rail. Roadbeds needed to be more solid. Bridges of wooden posts and sticks were once replaced with bridges of iron posts and rods. And then, those were replaced by massive bridges of steel. Tunnels have to be enlarged. Shops, engine houses, turntables and terminals have been rebuilt. You know, Ben, 1070 will exceed the clearances on many lines. And the new 2-10-4 will be too long for many of the turntables in use today.

We can't build locomotives any higher or wider. Locomotives have a maximum width, what we call the 'loading gauge.' If we build one wider than that, well, it'll get stuck in the first tunnel it comes to." Ben and Will laugh together at the image of a big fat locomotive stuck in a tunnel portal, its drivers spinning helplessly.

"There's just no end to the problems we engineers face," Will says with comical weariness. "If you add another set of drivers, the radius of some of the curves on the line will have to be increased. And big locomotives are hard on rails. I'll tell you, Ben, I've seen rails on the outside of a curve with heads worn down to a ragged knife edge, and the crews are replacing them every six months."

"But if you add more drivers, doesn't that take care of the problem by putting all that weight more evenly on the rail?"

"It does, Ben, and that's how engineers have attempted to solve the problem in the past. As fireboxes, boilers, cylinders, the whole locomotive got bigger and heavier, another pair of drivers was added—two, four, then six, then eight, then ten, twelve, sixteen. Take a look at this, Ben." Will opens up his latest issue of *Railway Age*. "The Virginian Railway is doubleheading 2-10-10-2s to haul coal from the mines in West Virginia to ports on the Atlantic seaboard. One of these monsters was at the head end of the heaviest train ever handled by one engine—110 cars weighing 17,250 tons. And look at this. The Erie has just put into service a 2-8-8-8-2 Mallet—count them Ben, twenty-four drivers—which weighs . . ."

"Look at that—853,050 pounds. That's... that's almost a million pounds. Unbelievable!"

"And so are the maintenance costs. I predict that they won't keep *those* locomotives very long. No, Ben, I'm convinced we have to find better ways to get more power out of a locomotive. We have to make the locomotive more efficient. We have to make a pound of steel and a pound of fuel do more work. That's what we have to do, Ben.

"I had a goal in mind when I started designing the 2-8-4. First, I wanted a locomotive that could pull trains ten percent longer than those the big Mikes can handle, at the same speed, burning the same amount of fuel. Second, I wanted to accomplish this with little or no increase in weight. With such a locomotive, 100-car trains could become 110-car trains with no increase in fuel costs and without slowing down. Imagine that." Ben could see that Mr. Woodard was enjoying this little lesson. "A ten percent increase in freight handling at *no* extra cost. Why, I'll tell you, that's no small thing these days.

"So, to begin. This is to be an eight-coupled locomotive. "I'll use 63-inch drivers. This diameter gives me the perfect combination of power and speed. The tall drivers you see on passenger locomotives—some are 90 inches in diameter—are for speed. With drivers like that a locomotive can do better than a hundred miles an hour. But it couldn't start up and pull a long, heavy freight train. Smaller diameter drivers turning faster will let the engine develop more horsepower. As with so many things in engineering, Ben, it's a trade-off—horsepower or speed.

"Up front here will be a two-wheel pilot truck. This leading truck supports the front end of the locomotive and l-e-e-eads it smoothly into and around curves. On the straightaway the pilot truck keeps the locomotive from twisting from side to side—we call it 'nosing.' You see, the pistons and rods on opposite sides work against each other; one is pulling while the other is pushing. That alternate pushing and pulling swings the locomotive first to one side of the track and then to the other.

"Those small wheels up front are important, Ben. They save lives." Mr. Woodard has suddenly become very serious. Ben can see the concern in his face. "Many years ago I was asked to help determine the cause of a dreadful accident on the New York Central. A locomotive had de-

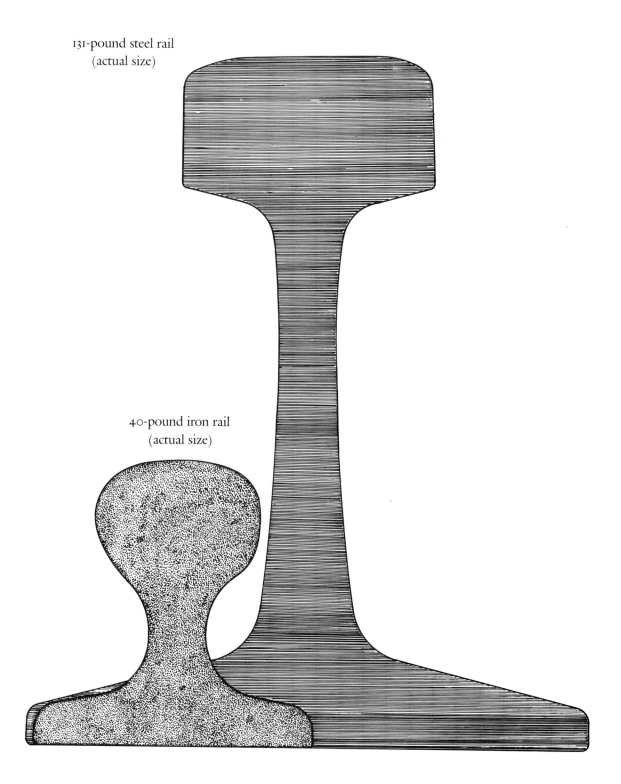

131-pound steel rail
(actual size)

40-pound iron rail
(actual size)

railed going into a curve at high speed. Many people were killed. Horrible, horrible thing it was. In the end, we still weren't sure what caused the accident, but I thought then—and I'm even more certain today—the culprit was a poorly designed lead truck. That locomotive climbed right up and over the rail on the outside of the curve. I'd bet on it. Since then, I've been very interested in pilot trucks, made a study of them. By the way, on high-speed passenger locomotives we usually use a four-wheel truck up front for added safety and stability.

"Well, to continue. We'll need cylinders and pistons to turn steam pressure into movement. And we'll need cross-heads in guides to support the piston rods and connect them to the main rods, like so. The force exerted by the main rod on the main drive wheel is called 'tractive effort.' You'll be hearing more about that in a minute.

"And, while we're at it, we'll complete the running gear with a Baker valve gear. The rotary motion of the crank is changed to reciprocating motion, sliding the valve back and forth, letting steam first in one end and then the other end of the cylinder. This way steam pushes the piston one way and then pushes it back the other—back and forth, back and forth, back and forth, turning the drive wheels.

"To increase tractive effort I need higher cylinder horse-power. So, I've increased the diameter of the cylinders to thirty inches, more piston area for the steam to push against. Just increasing the piston diameter a few inches can make a big difference. Mrs. Foster may have told you that doubling the diameter of a circle increases the area four times. 'Cylinder horsepower,' I should tell you, is a term I use. You won't hear other engineers use the term," Will flashes Ben a knowing smile, "but they will soon.

"Which gets us back to the firebox. If you were a fire-man, how would you make a hotter fire?"

"I guess I'd shovel faster, get more coal onto the grate."

"That would work . . ."

"Up to a point." Ben finishes Will's sentence and they laugh together. "'Up to a point' seems to be an important rule of engineering."

"You're on your way to becoming an engineer, Ben. As one of those firemen out there will tell you, there's a limit to how big a fire you can have. Throwing more coal in faster doesn't work after a while because the bed of coals becomes too deep. Air can't circulate through the thick, dense layer of coal and the fire smothers and smolders. So the engine smokes cindery black smoke. You waste fuel. Steam pressure begins to fall. And you've got what the fireman calls 'a wet mule in the firebox.' The problem gets worse on slow drag freights because there isn't enough draft blowing up through the grate to supply oxygen to the fire.

"No, just throwing in more coal won't work. The answer is a larger grate that spreads the fire over a bigger area. The grate area of the big Mikes is 66 square feet. But, according to my calculations, to get the steam I want for this locomotive I need a grate area of 100 square feet.

"Since I can't make the firebox any wider—loading gauge, remember, the engine stuck in the tunnel?—I have to make it longer. That means a lot of heavy firebox hang-ing behind the drivers. I can't rest this weight on the drivers because we've already reached the limit for four axles. And I can't put any more weight on a single-axle trailing truck. So, what do I do?"

"Add another axle to the trailing truck, and make it a 2-8-4, that's what."

"That's what I did. The firebox sits on two centering devices mounted on the rear of the truck. Rollers inside the pads let the truck swing around curves while still sup-porting the firebox . . . that's too hard to draw. Come on, Ben, I'll show you. The foreman of the truck shop sent me a note yesterday saying that they'd be mounting the pads on the new truck this morning. Let's go have a look."

Will gathers up his notes, grabs his suit coat from the oak rack and, with Ben almost running to keep up, heads out the office, shrugging on his jacket.

Will and his excited sidekick head for the truck shop talking as they go. A shortcut takes them through the block-long machine shop and out into the storage yard. Here are thousands of locomotive parts arranged like some giant's model train kit. Rough frame and cylinder castings await their turn in the machine shop. Wheels in piles or mounted on axles in long lines on tracks. Boilers primed with red lead on their way to the erecting shop. Pilot trucks, trailing trucks, smokebox fronts, stacks of piping, sets of drivers mounted on axles. Brake cylinders. Air compressors. Smokestacks. Domes. Brakeshoes. They con-

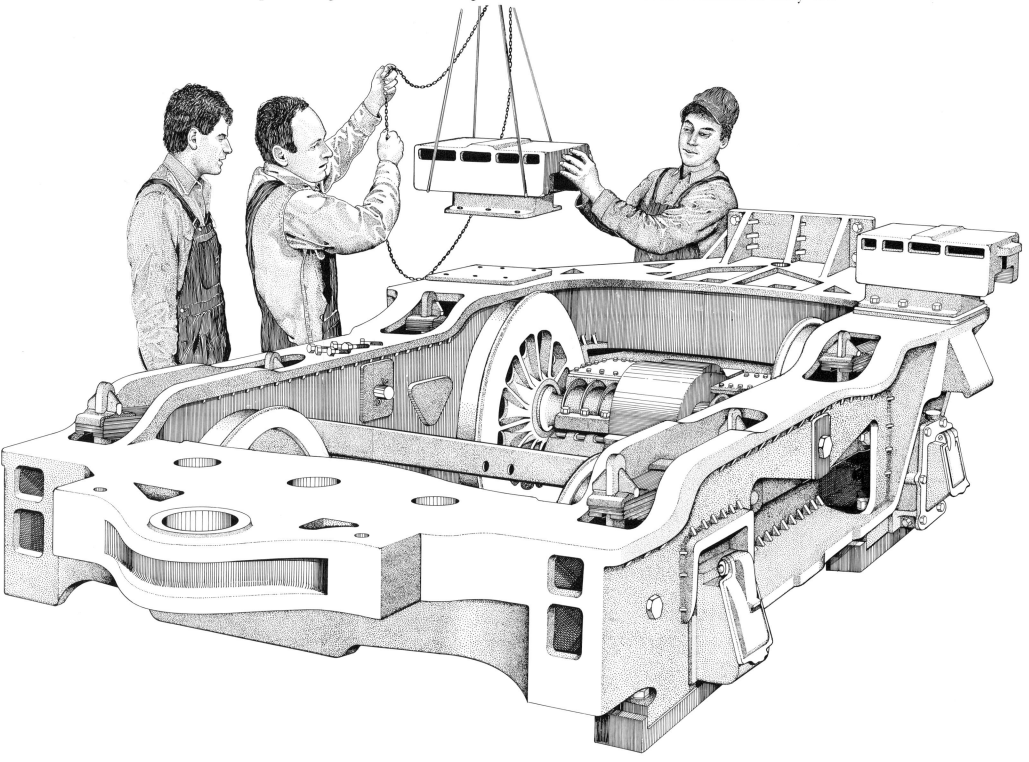

tinue past the smith and hammer shop and enter the truck shop building. And they arrive in time to watch the pads that will support No. 1070's big firebox being bolted to the new four-wheel truck. "A few weeks from now, Ben, in the erecting shop, you'll see just how nicely those pads work."

Ben and Will find a drawing-strewn table in a quiet corner of the shop to continue their lesson. "I have another way to increase horsepower," Will continues, spreading his and Ben's notes and sketches out over the blueprints. "I can make steam work harder by making it hotter, by superheating it. When I was an engineering student I learned a rule of thermodynamics that says: the hotter the steam, the more cubic volume it expands to fill. In a confined space—like a locomotive cylinder—superheated steam exerts more pressure against the piston, works harder. The superheater is here, in the boiler.

"Do you remember the temperature at which water boils, Ben?" The question takes Ben back a couple of years to Mr. Crane's science class and the rattling lid of the teapot demonstrating steam pressure. "I think it's 212 degrees."

"Right. But the steam coming out of 1070's superheaters will be over seven hundred degrees. And it makes quite a difference.

"That trailing truck, you noticed, not only supports the larger firebox but it also has a booster engine in it. Well, that's another little trick I learned designing the earlier Mikes. A booster's just a little steam engine, but it really works. I figure the tractive effort of the locomotive to be 69,400 pounds. The booster engine raises that to 82,600 pounds. That's a real boost I'd say.

"Well, Ben, there you have it. All the essential workings of a modern day steam locomotive, things it took me years

to learn, all in a morning's lecture. It probably looks to you like everything just works out, but I'll tell you Ben, I've made a lot of mistakes along the way. Have to. Making mistakes is how we learn. It may sound strange, Ben, but I hope you make a lot of mistakes in your career. It's natural for a young man to be striving for success all the time—you'll be good at whatever you do. But what's really important, son, is your attitude toward failure. If you've never failed, well, then it means you're not taking any risks, really stretching. And if you don't take any risks, you'll never learn any more than you already know.

"Well, enough lecturing. Now sir, how many of these new 2-8-4s would you like?" Mr. Woodard reminds Ben of the salesman down at Dave Weil's Haberdashery. "I can have Mr. Erickson write up the order and contract for you right now."

"I don't know." Ben's brow is furrowed in seriousness. "Probably just one. I'd have to ask my vice president in charge of operations. They're pretty big, you know. I'd have to find a place for them, and Mom's not very happy about the condition of my room as it is. Besides, I'll bet a locomotive like this is awfully expensive."

"Well, let's just see." Will takes up his slide rule, does some calculations and jots down a few figures. It all takes about half a minute. "You can have a shiny new 2-8-4, painted in a nice shade of black with white wheel rims and handrails, the numbers on the cab and your road's name on the tender in aluminum leaf, leather-upholstered cab seats, polished bronze bell and . . . oh yes, a complete set of tools, ready to run for $184,140.00 and a few cents."

"How did you do that so fast?" Ben wonders if Mr. Woodard isn't actually a magician.

"It's really quite simple, Ben. Locomotives, you see, are like potatoes and beans—you price them by the pound."

◇ 4 ◇

Foundry

BEN STANDS at the foundry's gaping door with the sunshine on his back, looking in on another world. Inside it is dark and gray. Even the air is gray, clinging to his face and hair like a smoky curtain. His throat parches and he can taste the thick air bitter on his tongue and gritty in his teeth like grains of sand.

He is fascinated. The floor—if you could call it a floor—is an unearthly landscape of black sand. Ben reaches down to pick up a handful and lets the fine, oily grains sift through his fingers. From somewhere way back in the grayness comes the roar and heat of immense furnaces. Here and there, in the swirling eddies of dust, Ben catches glimpses of dim figures moving about like phantoms in a fog and flashes of green-yellow fire.

The mystery draws Ben in. He steps warily onto the sand and makes his way toward a group of workers in the gray distance. Light rays of sparkling crystal dust fall from the clerestory windows high above.

What happens next, happens too fast to tell. Ben only remembers his right foot sinking into the soft sand, to the top of his boot, and his foot getting very hot, burning. His leg jerks upward and he sees that the leather is scorched and his laces are burning. He dances around madly, kicking into the sand, trying to put out his blazing laces. Suddenly, from behind, powerful hands grip him by the collar and the seat of his pants, fling him through the air and splash him feet first into a concrete tub. Ben stands dazed, the

cold water up to his knees, looking at a circle of laughing, black-smudged faces like apparitions in the dusty shadows.

"Hey, you okay?" a kindly voice from the circle inquires in a guttural tone.

"Fine . . . Okay . . . I think," Ben answers, not yet sure and feeling a bit silly about his predicament. His foot isn't hot anymore, but the water has already soaked up to his waist and his shirttails are very cold. The moment reminds him of some episode out of Krazy Kat comics.

"You are a lucky one." A powerful, dark man steps forward, stretches his hand out to Ben and pulls him from the tub. "Have a hot foot, no?" Ben looks into black eyes hard and shiny like pebbles under bushy brows. The friendly, wrinkled face is set into a ball of curly hair and a gold tooth gleams in his smile. From the size of the foundryman, his short thick arms, Ben figures he must have been the one who picked him up. "So now you are a foundryman. By golly, you a real foundryman. We have all done that. We all step in the iron."

The molders and founders gather around Ben, slapping him on the shoulders, jostling him. There's good-natured laughing all around. One wears an old slouch hat, another a leather apron, all have great mustachios. Ben gathers they speak little or no English. But in their throaty voices he recognizes the accents and lilt of the girl who sat next to him in class from a far-away place called Serbia. Just a few minutes ago he was a stranger. But now he feels welcome

474-A-29

1070

as though he had come through some sort of initiation and is now worthy of their brotherhood.

Ben's rescuer explains what happened. "Sometime we finish pouring and is still iron in the ladle. So we dig a trench here. And into the trench we pour the iron. That iron is called—how do you say that—a 'pig.' When the pig is cold we break it up and put it in scrap pile for next time. I cover hot pig with sand and you step in it. But you are okay. And we are still friends. My name is Marko, Marko Ukropina."

Ben explains to Marko that he saw a schedule in Mr. Schnell's office, and he figured the foundry would be working on Order No. 1070 about now. "I'm sure . . . um . . . interested in that locomotive"—Ben was finding it a bit hard to get his personal feelings about "his" locomotive out into the open—"I'd sure like to follow it along. I guess that's as good a way as any to learn about locomotive building."

Marko agrees, but suggests to Ben that if he really wants to follow 1070 from the very beginning, he needs first to visit the pattern shop. "You go talk to the Cunnan brothers over there. Then you come back here, Ben, and I show you how we make the locomotive."

The pattern shop is as bright as the foundry was gray. The light wooden floors and bench tops, the fine dusting of white sawdust everywhere, remind Ben of his sixth grade wood shop class. The smells are familiar too, of pine, fir, and oak—sweet, pungent, spicy. And the sounds, the whine of table saws, tap-tap-tap of wooden mallets against chisels, rasping files and sandpaper whispers. The wood dust tickles his nose. Apple-red patterns stand among the benches, saws and joiners and in stacks along the walls. Ben recognizes among them beautifully crafted wooden drive wheels, smokestacks, a bell, a pilot beam and other locomotive parts, some with "1070" stenciled on them.

Iron, steel and brass castings, some weighing thousands of pounds, begin as fine, crafted wooden sculptures called "patterns." A pattern will be made for each of the hundreds of castings needed to construct the new 2-8-4. The smallest castings, like brass handles for valves, will weigh only a few ounces. The largest, the steel side frames, will be over forty feet long and will each weigh more than 20,000 pounds.

Ben arrives in time to watch the making of two of 1070's patterns. John Cunnan is just putting the finishing touches on the pattern for a footplate, part of Will Woodard's new four-wheel trailing truck. Like most of the patterns he builds, this one is made of sugar pine and some maple and different hardwoods.

John's brother, Robert, is checking the dimensions of the pattern for 1070's front deck casting (which you can see finished and bolted in place between the main frames on page 91). When he's checked all the measurements against the blueprint, and is certain that the pattern is correct, Robert will have one of the apprentice pattern makers sandpaper it to silky smoothness and then paint it bright red.

Robert explains to Ben that the pattern isn't made to the exact size shown on the drawing. The patternmaker has to keep in mind two things that will happen to the casting made from his pattern. First of all, the casting will be smaller than the impression left by the pattern in the molding sand. That's because the molten iron or steel poured into the mold shrinks as it cools and hardens. So if Robert wants the casting to be the same size as the drawing calls for, he must make the pattern larger than that. "I don't have to guess," he explains to Ben, "because I know how much different kinds of metal shrink. I know,

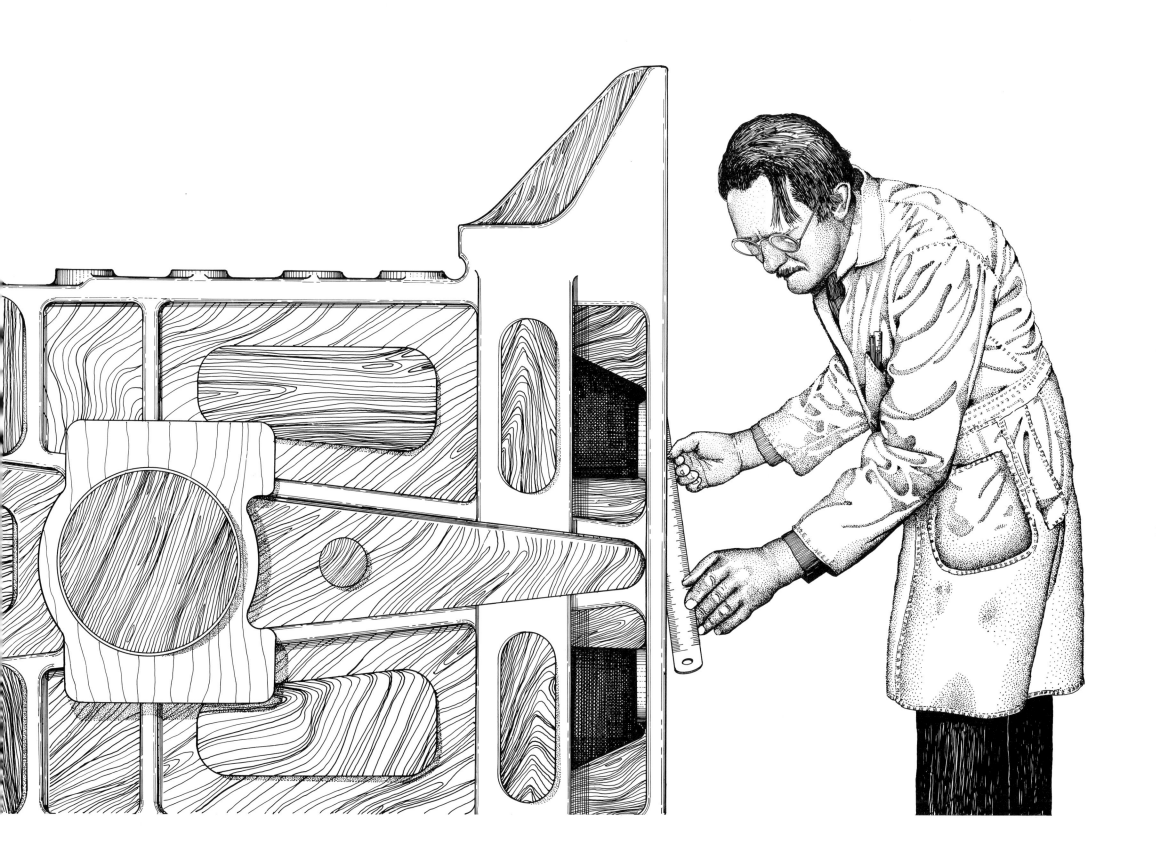

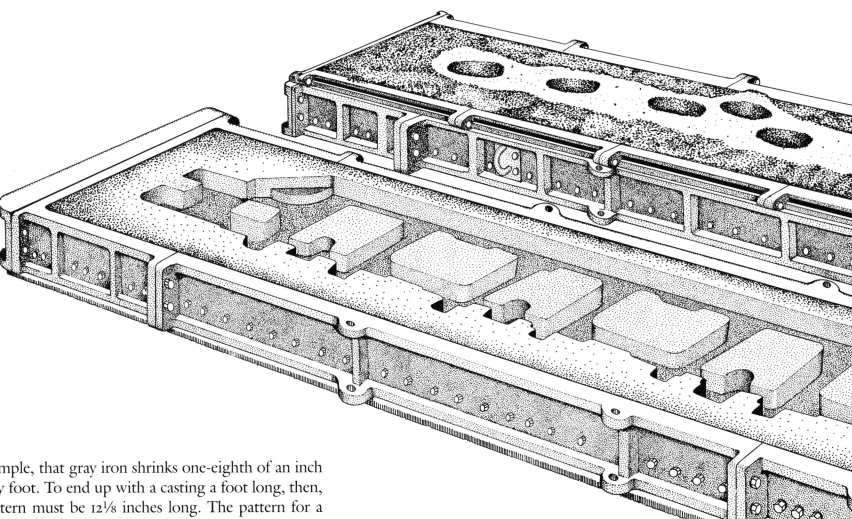

for example, that gray iron shrinks one-eighth of an inch in every foot. To end up with a casting a foot long, then, my pattern must be 12⅛ inches long. The pattern for a two-foot-long casting will need to be 24¼ inches long and so on. I use this special scale called a 'common rule' to lay out my patterns. The inches and feet on this rule are 'stretched out' so that whatever measurement I take with it, the shrinkage is figured in."

On the drawing Robert shows Ben another reason for making the pattern larger. "Several places. . . here . . and here . . . and here, the engineer calls for a finish on the rough casting. One part might be finished smooth on a planer, another on a miller, or lathe or grinder. Whenever you machine a casting to give it a smooth finish and get it down to an exact size you're taking metal off. I have to figure in how much metal the machinist will have to take off. Then I have to put a little extra metal on each of these

places. If I don't, the finished piece will end up being too small.

"There are other things I need to think about, like what we call 'draft.' But you'll see that more clearly when you've spent some time in the foundry. And watch out over there; if you stay too long Marko will make a molder out of you."

Ben returns to the foundry a hotfoot wiser. His new strategy for staying out of trouble, as he searches for Marko, is to keep along the edges of the black desert. He

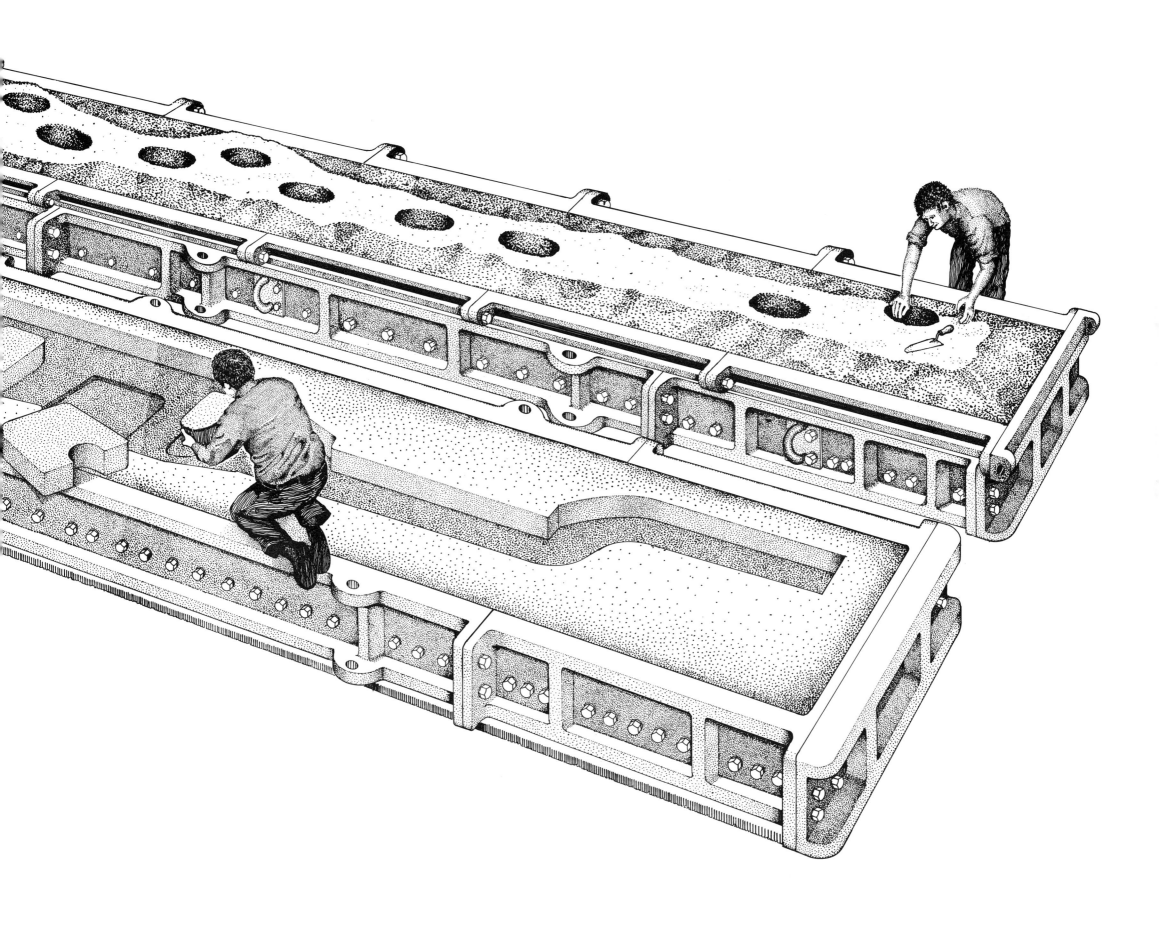

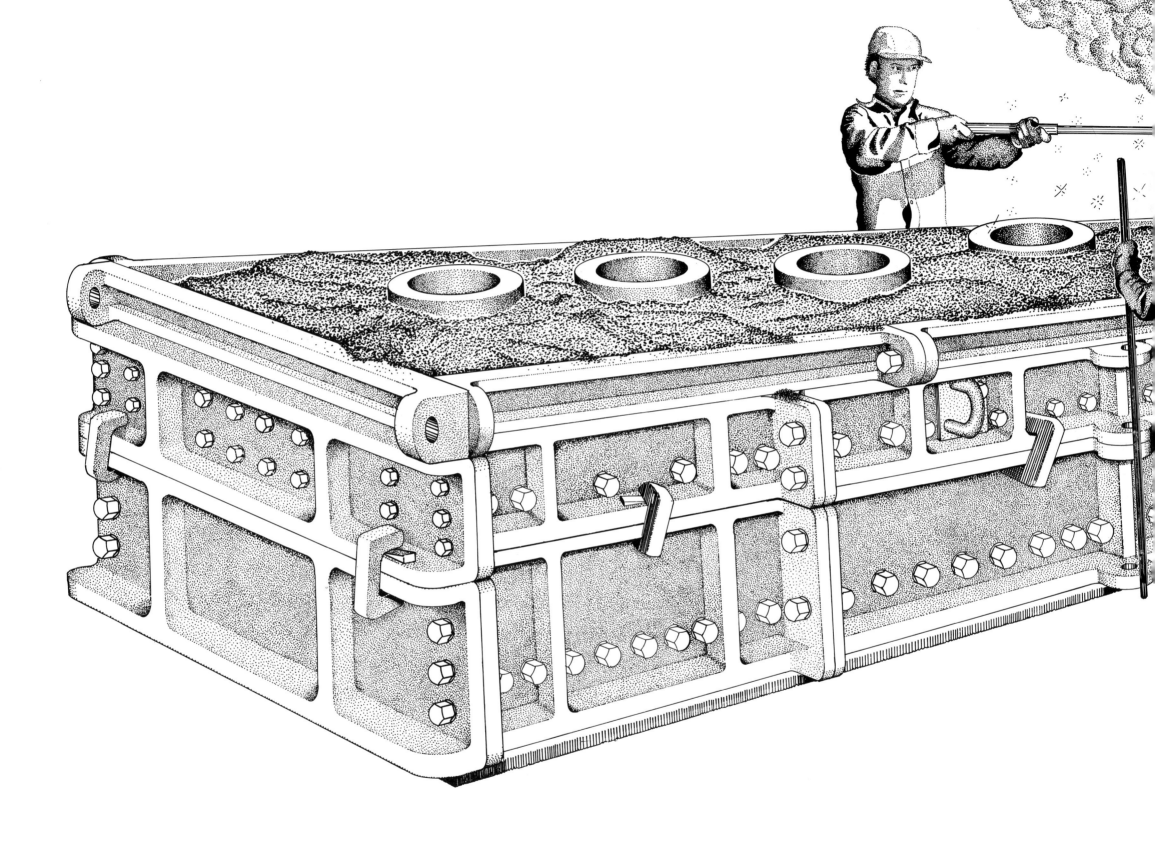

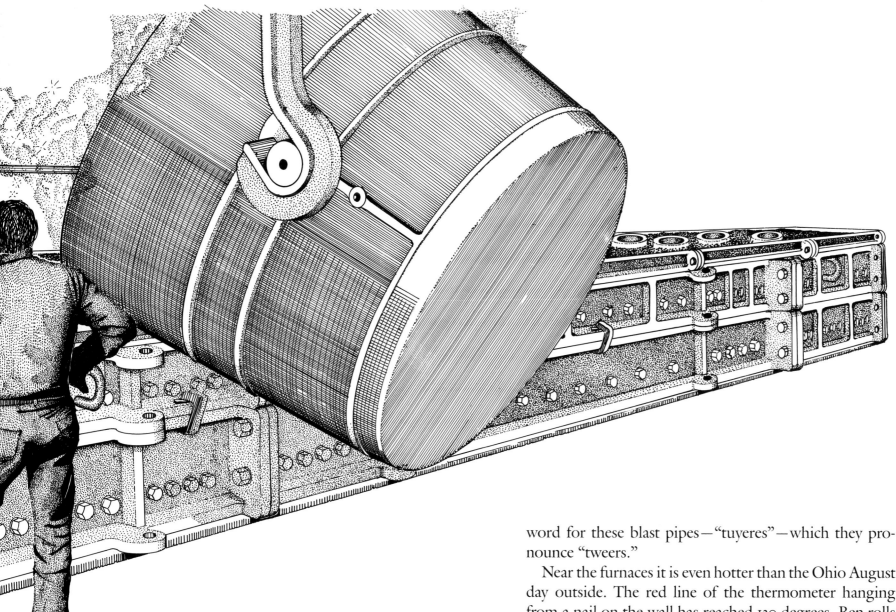

discovers that the pulsing roar that fills the shop comes from the tall stack-like cupola furnaces made of riveted, overlapping iron plates, ascending high up through the roof. Inside the cupola walls, lined with fire bricks, pigs and scrap iron are melted to liquid metal in layers of red-hot coke. The roar is the blast—hot air pumped through pipes into each furnace to force oxygen up through the layers of burning coke and metal making them still hotter. Ben discovers that the foundrymen still use an old French word for these blast pipes—"tuyeres"—which they pronounce "tweers."

Near the furnaces it is even hotter than the Ohio August day outside. The red line of the thermometer hanging from a nail on the wall has reached 130 degrees. Ben rolls up his sleeves and pulls the bandanna from his back pocket to mop his neck and face. The stolid faces and arms of the olive-skinned men around him glisten with sweat. Ben ties his drenched bandanna around his neck.

The cupola man seems, to Ben, to be peering into the roaring hot furnace through a telescope. "Is 2,500 degrees in there now," he calls out to Ben over the roar of the furnace blast, beckoning him with a gloved hand to come closer. "See, you look in there." Ben puts his eye to the little mica lens and looks right into the furnace. Inside are glowing red rocks. "You look and you see iron running there," the cupola man shouts close to Ben's ear. This time Ben notices the red-orange molten iron trickling down through the burning rocks. "Is coke . . . made by roasting coal. But is burning hotter than coal."

Michailo Milojevich tends his furnace with a patience and skill he learned in the old country. He came to America in 1895, first to Ellis Island and then here, to Lima, where his brother had settled before him. He was one of tens of thousands who came from Italy, Greece, Roumania, Hungary, Czechoslovakia, Serbia and Croatia, Poland, Latvia, Estonia, Lithuania, from all over southern and eastern Europe that year. They hadn't come to build locomotives—most had never seen one until they left their villages for the New World. But they brought with them skills, traditions of work and muscular bodies that suited them well to building these powerful machines.

Michailo is a master foundryman who takes pride in the quality of his iron. As he tells Ben, he's been around men working iron for as long as he can remember. "I was a young kid, like you. In my village a young kid do this, do that. Little by little I am learning how to work the iron." The temperature of his cupola must be exact—too cold

and the iron will not flow, too hot and it will become brittle and useless. He has no thermometers, no gages to tell him how hot it is inside. He knows from the color of the coke and molten iron, from how easily or sluggishly it runs, from the sweat that rolls off his body.

◇

Ben finds Marko working on an immense open sand mold for one of 1070's side frames. The upper part, or cope, has been lifted off and set down beside the lower part, the drag. "The sand. It must be right," the foundryman explains as he works. "Too little water is not good. The sand is not holding together. Then when you pull the pattern, the mold fall in. If the sand is too much damp then that is not good also. You pour in the hot iron or steel and it makes steam. Bang! The mold blows up. It must be just right, Ben, not too much damp, not too little damp. The foundryman knows when the sand is good. He is feeling it in his hands. There, it is time to take out the pattern."

Marko and the craneman high above work together through hand signals and hardly a word. The long wooden pattern is painstakingly lifted up and out of the sand with

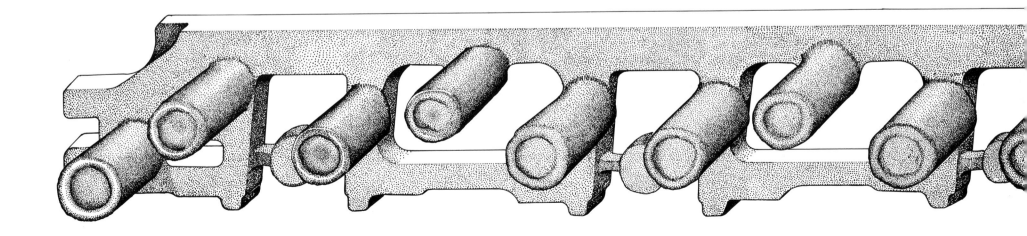

the overhead crane, leaving its impression in the sand. Ben examines the pattern as it hangs from the hoisting block and notices that the sides taper downward, into the mold. "There it is," he says to himself, pleased by his discovery. The tapered sides let the pattern be lifted up and out easily, without breaking the mold and the smooth sand surfaces. This is what the patternmaker meant by "draft." Ben's probably not aware of it, but he is learning to see details of the work he would never have noticed before.

Marko positions cores—made of special sand formed in wooden molds and then baked hard in an oven—where they will leave openings in the frame. Each core fits into a core print, a shallow impression left in the sand by the pattern, so that the cores won't move when the molten steel rushes into the mold.

After the cores have been set, Rade Vuckovich finishes his work on the risers in the cope. He is Marko's cousin and only last year came from the village where they grew up together. Rade speaks little English but it is not important in a shop with so many of his countrymen. Again the overhead crane is signaled for. The chains are made fast and in a whir of powerful electric motors the huge cope rises slowly into the air. Guided now by a dozen powerful

arms, the cope is swung over and lowered—careful now—squarely onto the drag. Inside remains the hollow space left by the pattern that, when filled with molten steel, will become a locomotive frame. Finally, the cope and drag are dried with gas-fired torches and readied for the pour.

The time for the pour has arrived and the founders take up their positions. From out of the hazy distance comes the great ladle aglow with a fiery charge of twenty thousand pounds of vanadium steel. Marko signals the crane operator and the ladle descends slowly to the mold. The craneman inches the massive ladle back and forth until the pouring beak is in just the right position. Another signal from Marko. Slowly the ladle tilts down until a syrupy steel bubble rolls over the edge and pours in a tapering stream down into the gate, the opening into the mold. The foundry brightens as though it is on fire. Faces and blue clothes whiten in the brilliant glow. As Marko guides the craneman, another founder holds back the hissing, sputtering slag—bits of coke floating on the surface of the molten steel—with a long, iron skimmer. Fourth of July sparkles flash through the air. As steel fills the mold, burning gases vent up through the risers, erupting into flickering green flames.

Another signal from Marko brings the second ladle which begins pouring just as the last drops of steel fall from the emptying beak. In a few minutes Marko sees the telltale sign of the pour's end. Yellow-orange steel has filled each riser to the brim. Marko's arm shoots up into the air and waves back and forth, and instantly the ladle tilts up, stopping the steel flow. The ladle ascends and glides away as mysteriously as it appeared toward the furnaces for another charge. When it returns it will be to pour the second frame mold.

After a few hours Ben returns to find that the molds have been broken and the two frames, now dull cherry-red, left to cool. Marko takes Ben over to look at the new frames for 1070. Ben's face and hands burn with their heat. "Now you will see how the risers work. The mold is filling up with steel, yes? So the gases come up through the risers. When the risers fill with the molten steel I know the mold is full and the pour it is finish." The gate and risers will be cut off tomorrow and the frames moved to the annealing oven.

The side frames are the foundation of the locomotive. The firebox, boiler and cylinders are fastened to the frame which bears and distributes all this weight to the drivers. All of the forces of the locomotive—the jarring and jouncing of uneven track, the hammering of the drivers, the jolting pull of the train behind, the twisting on sharp curves, the wrenching of the cylinders pushing and pulling, the vibrating, shaking, bumping, lurching of hundreds of thousands of pounds of speeding iron and steel—must be absorbed by the frame.

So the frame must be rugged and resilient. But, right out of the mold, frames have cast-in weaknesses—what engineers and founders call stresses. This happens because different parts of the casting are different thicknesses, and the thinner parts cool faster. The unequal cooling causes unequal shrinking and, whenever that happens, there is stress.

Before the frame castings can be used, then, the stresses will have to be relaxed. And to even them out, Marko's gang will place the castings in a cavernous annealing oven. Here the castings will be heated up to 1,600 degrees. Then the heat in the oven will be turned down gradually until the oven and the castings are cool. Once again, the castings are heated and slowly cooled so that all the parts of the frame, thick and thin, cool down together.

During his visits over the next few days, Ben watches other castings of iron and steel emerge from the foundry's sand. The delicate white lines of Will Woodard's blueprints are being transformed into massive forms of gray metal. But Will's and Ben's locomotive is still rough. When Ben follows the castings over to the machine shop, he'll see them worked into gleaming, precision shapes.

◇ 5 ◇

Forge

BEN ALWAYS knows just where to find his grandfather when he comes to the hammer shop. He'll be at his forge wearing his heavy leather apron, hammering a piece of steel on the anvil with the vigor and force of a man half his age. At first he wouldn't see Ben standing in the corner out of the way of the sparks. Then he'd catch sight of his grandson over the top of his glasses and lift his hammer to the boy in greeting. Ben's favorite image of his Grampa Joe is with a rod in his hand, hammering with big swings of his heavy maul, sparks flying around his head, his whole body coiling and uncoiling like a powerful spring.

Joseph has been here longer than most men at The Loco can remember. He was Ben's age when he left Finland and came to work at what was then the Lima Machine Works. That would be about 1880. Actually, he's certain when he began work but he's not sure how old he was then—or is now, for that matter. The way he explains it, people were too busy in his little village of Neibatt to spend much time keeping records. When his young grandson would become impatient with him because he didn't even know when his birthday was, he would always laugh and answer, "Ya, well I was born and I am here, and dat is how Benny gets marbles and a taffy apple on Sundays."

Ben's first hand tools were the carefully crafted hammers, awls, augers, planes, punches, scales, calipers and chisels his grampa had made and brought with him to America. That was one of the treats he looked forward to as a kid. Grampa Joe would put him up on the heavy beechwood workbench which smelled of linseed oil (and which Grampa Joe had made), right in front of the tool chest.

Ben recalls even now what fun it was to open the little felt-lined drawers and take out each tool. Early on, Grampa Joe had taught him how to handle tools, to respect and care for their smoothly lapped surfaces and sharp edges. When he was older and could actually use the tools, he would imitate his grandfather's way of wiping off each one with an oily cloth before putting it away.

Grampa Joe's tools would always be for Ben what craftmanship should be. They remained a standard by which he measured everything handmade. But there was a special warmth in the soft patina of the metal and in the grain of the wood he didn't feel in others. A few tools did seem to Ben less exact somehow. These, his grandfather explained, were the first ones he made, when he was ten or eleven. They were not, to the master's eye, well proportioned or quite balanced for the hand, but he had made them with all the love and care he could and so he had kept them.

The old man liked to tell his grandson, as he had his son years ago, stories that were fun to tell but also held some valuable lesson. Patience and careful work were Grampa Joe's favorite themes. Ben's visits with his grandfather at the forge or working together at home in the workshop were usually the occasions for these stories. Some of them became such favorites that Ben would ask his Grampa Joe to tell them again and again. Once, when the old craftsman was showing the boy how to smooth the hammer head he was having him make with a file, he told him about his own introduction to using a file longer ago than Ben could even imagine.

One day, when he was eight or nine and getting into a lot of mischief in the smithy, his father sat him down at the bench with a chunk of steel—a small piece about the size of a matchbox—and a file that he had made. His task, his father told him, was to get the piece of steel "rectangular, square and smooth." He'd work at it for a time and show it to the stern old blacksmith, who would lift his glasses onto his forehead, hold the little block of steel close to his eyes and rub it softly with his fingers. Each time he would return it to the boy, whose eyes were filled with expectations of praise, and say gently, "rectangular, square and smooth, Joseph." So it went, again and again. Each time the boy thought he had completed his task to perfection, and searched his father's face for some glimmer of approval, only to have the block put back into his hand with the same intoned instructions, "Rectangular, square and smooth."

"The more I'm workin on, lookin at, touchin the little piece of steel, the more I'm understanding the meaning of my father's words—'rectangular, square and smooth.' Remember, Benny, we have no machine tools in our village. We have not even one of the wonderful millers and planers you see today. Ya, so the finest, smoothest finishes we made with our hands. Now you will not believe this that I tell you," the old smith warned his grandson. "Before that piece of steel was right to your great-grandfather, I had filed and filed for nearly a year, ya, a year."

When Joseph came to work here many of the buildings

hadn't been built yet. Locomotives were smaller, less complicated then. So was Lima and, to his way of thinking, so was life. A lot more of a locomotive was made in the smithy and forge in those days. Over the years Joseph has made many parts of a locomotive, all by hand.

The forge is a special place for Ben. His grandfather is here. And he likes the work of the hammer shop. For Ben, the work of the forge is the most straightforward, most spectacular work of all. This is where the working, moving parts of the locomotive—the side rods and connecting rods, valve gear, the axles, cranks and crank pins, even the tires—are hammered with enormous force out of billets of red-hot steel.

There's wonderful action here. Heating furnaces rage with a roar—fierce, gaping 2,400-degree infernos. Bolt headers chatter away forming red-hot round bars into rough bolts with round, hexagonal or square heads. The tup of a small drop hammer strikes the anvil forty times a minute, forming a perfect jacket lug each time. Powerful shears slice through twelve-inch steel bars with the ease and speed of a wire cutter. Sizzling steam rises in clouds from the quenching of hot steel in water. The tup of a massive, squat steam hammer plummets to the anvil delivering a groundshaking blow. *Thud, thud, thud, thud,* Ben can feel the tup striking tons of incandescent steel through his whole body. Ben has never felt an earthquake, but whenever he reads about them he imagines they must feel like this.

It is hot and when the billets emerge from the furnaces, the forge lights up with the glow of a hundred electric lights. But what pleases Ben most are the gangs of bull-necked hammermen leaning into the tons of steel suspended on stout chains. From furnace to anvil to furnace, back and forth they go, heels dug in, muscles straining, locked together in a dance with the great hunk of glowing metal. The antics of the barrel-chested hammermen make him laugh, reminding him of the tugs-of-war at Sunday picnics (and now he knows why the forge team always wins).

No matter where you are at The Loco, the 24,000 pound blows of the big hammers can be felt. Ben has heard his dad and Joe Park and the other machinists talk about what happens when the hammermen start roughing out an axle or side rod. Lathe, planer and shaper cutters jump and skip with each blow. Working to close tolerances becomes unthinkable. Only rough cuts can be taken or the machines must be shut down for a while. "When Ed Rutsch and his gang get that big hammer agoin, there's no finish work to be done around here," the machinists grump. The anvil of a double-frame steam hammer sits on a foundation buried deep under the woodblock floor, but the impact of each blow travels far and wide. "The whole of South Lima shakes every time that forge hammer starts up," folks often say, whenever the weather's not worth talking about.

The hammer shop has another distinction. It's where Ben's dad, Grampa Joe and a lot of the other men get their laundry done. Long ago, long before anyone around except Grampa Joe can remember, Old Fess, who tends the boilers that make steam for the hammers, discovered that the chemicals he puts into the boilers to keep them clean also clean greasy overalls. So, for a dime, Fess will put your blue denim overalls in a vat of boiling chemicals. Ben could always tell who's had Fess wash their overalls. They come out clean as a whistle and bone white.

Ben goes to watch his grandfather work at his forge. The forge is heaped with glowing coke. A blower forces air up through the hot coals making them glow even redder and hotter. To begin, he puts the end of the rod he wants to work into the coals. When the rod reaches red heat, he takes it from the fire and hammers it to shape on his anvil.

The smith can work only a short while, until the piece begins to cool, before having to return it to the fire. The iron is heated until it is workable again and then taken to the anvil. Grampa Joe knows when the iron has reached working temperature by the color. Dull red iron is still too cool to work. Bright cherry red is just right. The smith has special hammers to help him form different shapes, and punches to punch holes through the soft iron. And by hammering two heated ends together he can make a weld.

To Ben, the remarkable thing is how workable the metal becomes. It seems impossible that hard, heavy iron could become soft. But with a hammer and a few simple tools, the smith can fashion a bar into a hammer head, a prybar, a wrench, cold chisels, punches. Joe and his gang make the tools that go into the tool box in the cab of each locomotive—coal pick, fire hook, hammers, grease cup wrenches—and many of the hand tools used in the shop. Not too many years ago, these smiths hammered out ornamental ironwork for locomotives like headlight and bell brackets, done up in lacy iron filigree. Grampa Joe misses those decorative touches. To him they recall another age. "We made everything by hand in those days, Benny," he's told his grandson. "They were something to see, those little locomotives, and as good as they make 'em." But he's also a practical man and knows about change.

Ben knows the locomotives his grandfather first built only from pictures. But he does remember his grandfather making replacement parts for his old tractor and the Aeromotor Chicago windmill that pumps water on the farm. Grampa Joe's small forge out in the barn is a link between his life and his work, making them one.

The work of the smith and the hammerman are the same. It's just that they work on a different scale. Instead of small rods and bars, the hammerman begins with a billet of steel, heats it in a large furnace and then hammers it into the shape he wants, not with a smith's maul but a big steam hammer. A side rod for Ben's locomotive begins as

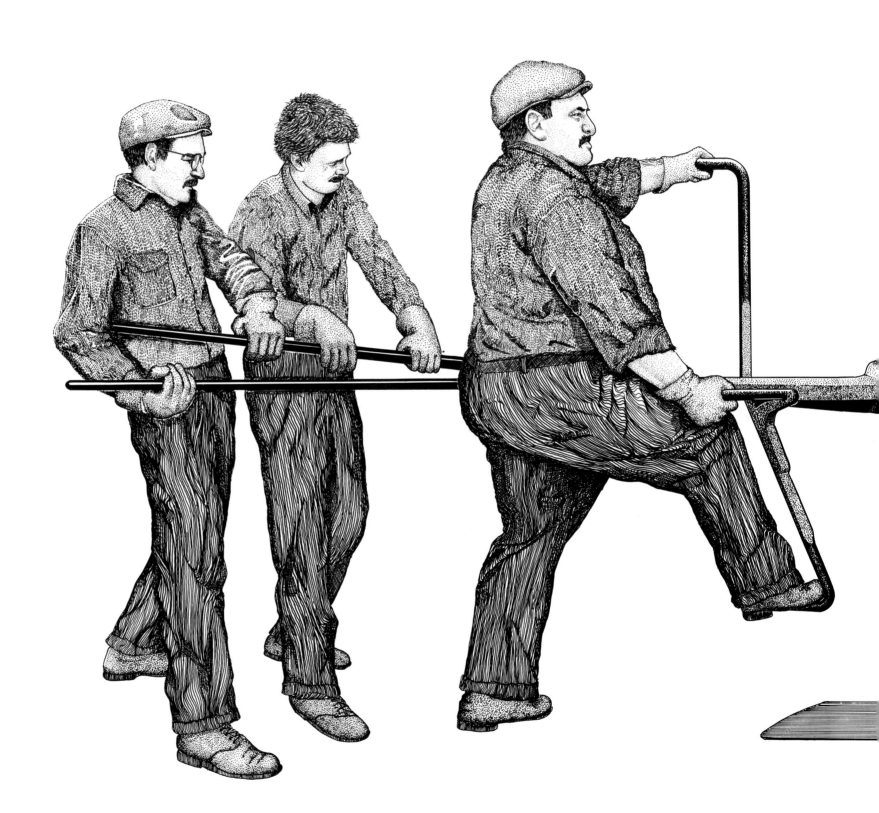

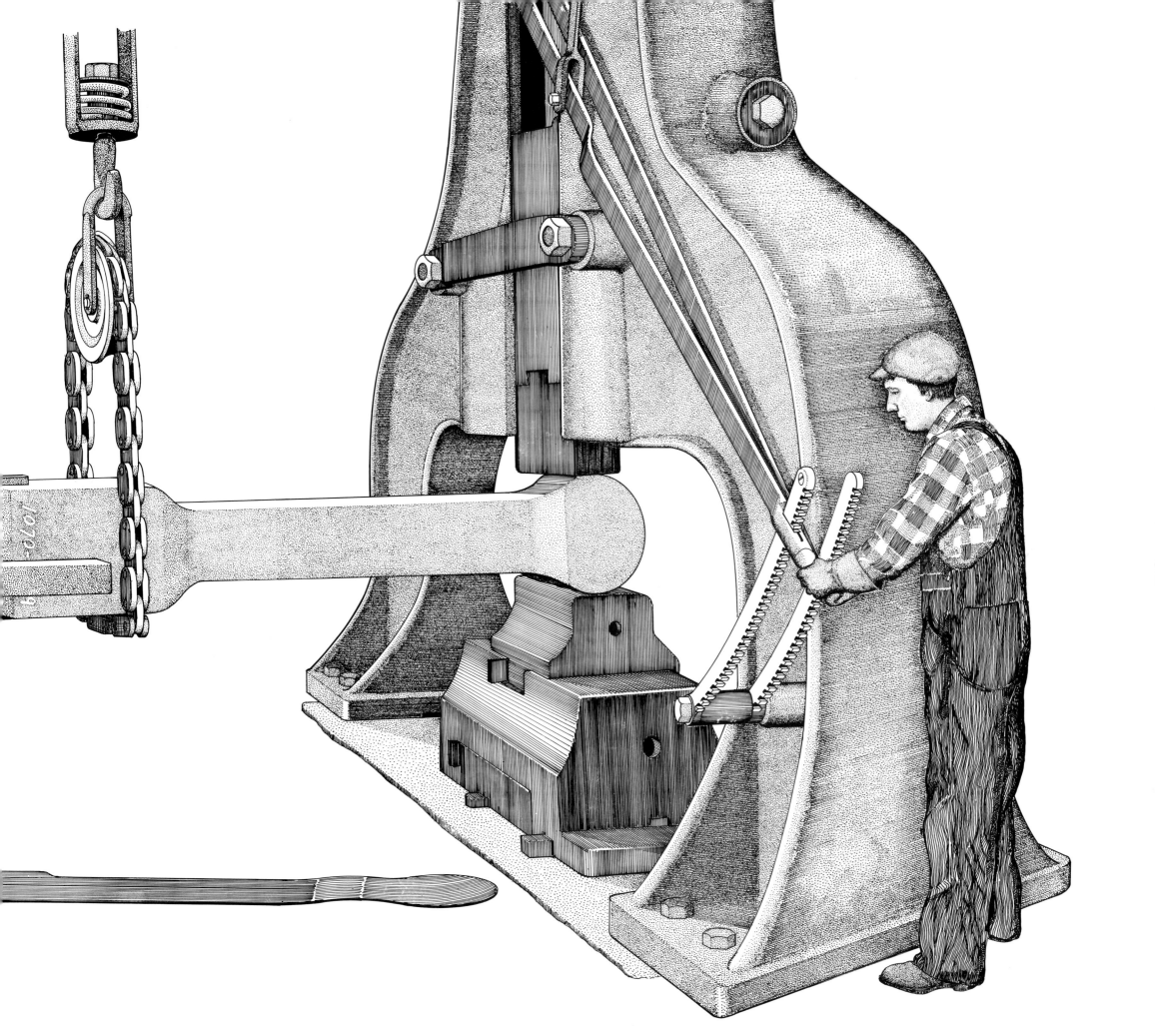

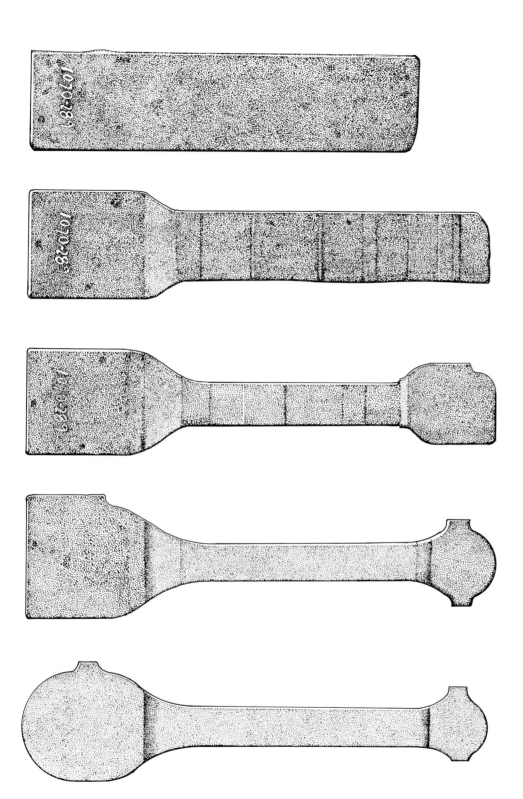

a steel billet weighing 3,600 pounds. It is forged into shape by a double-frame steam hammer that can strike with a force of thousands of pounds.

Not all the steel being worked in the forge is the same. Different parts of the locomotive call for different kinds of steel. Just what kind of steel the engineer chooses depends on the work the part will be doing and the kinds of forces it will have to endure. Steel is given different qualities by adding different elements to it. It is then called steel alloy. Adding carbon, for instance, makes steel harder, more resistant to wear. Adding exact amounts of manganese or tungsten makes the steel even harder. That's why the tools used to cut iron and steel on the lathe, milling machine, planer and other machine tools are made of carbon or tungsten steel. Carbon and tungsten steels can be made so hard that they cannot be machined by cutting.

Engineers are always having to make choices about the materials they use. High-carbon steel is so hard that it cuts or drills easily through other steels, but it is brittle. Mild steel bends and flexes without breaking, but it wears quickly. So in his plans for the 1070, Will Woodard specifies a "mild steel" casting for the trailing truck frame. It will have to support the weight of the firebox and withstand a lot of twisting and flexing, but it needn't be hard against wear.

The work of the axles and piston rods is different. These, Will directs the hammermen, are to be forged out of a billet of "medium carbon steel." High carbon steel is harder and would wear even better, but Will knows that the forces acting on the axles and piston rods would shatter high carbon steel.

Sometimes the steel in a locomotive must do everything well, resist wear and withstand tremendous twisting forces. That's what crank pins have to do. The crank pins receive the full power of the cylinders through the main rods, and transmit it to all the drivers. (There's a crank pin in Joe Park's lathe on page 66.) For the crank pins, then, Will Woodard specifies "chrome vanadium steel." The little bit of chromium added to the steel makes it hard against wear. An even smaller amount of vanadium increases the crank pins' resistance to shock and the stresses of the pushing and pulling of the main rods. When 1070 starts up with a mile-long string of freight cars behind it, the total tractive effort—69,400 pounds—will act against the crank pins, two tough steel cylinders 22 inches long by only 8 inches in diameter.

Ben moves up behind the hammerman operating one of the double-frame steam hammers. What intrigues him is the control the hammerman has over the ram. With an operating lever in each hand, he can move the ram up and down a few inches, lightly tapping the forging, or raise it the full height and bring it down in an earthshaking blow. As Ben watches, a glowing red billet of steel becomes a side rod for his locomotive.

The billet has been at soaking heat in the furnace for several hours. As the forgemen wait, Jan Hovorka motions to Ben to come over by the furnace. Like most of the crew he is a big, brawny man. He wears a hoary old woolen shirt, frayed suspenders and thick leather apron. "So you want to be a smith." Jan has to shout over the furnace blast and the pounding and clattering of a dozen small and large hammers. "Well, I have heard your grandfather talk about you since you are this high. You have grown big since then. I think you are ready for lesson."

"First we are heating the billet. We get that from a steel mill. It has been heating at 2,280 degrees. It will come out very soon now. The temperature must be right. If it is getting too hot we call that 'burned steel.' That steel could break and fly apart when we hammer it. If the steel is getting too cold that is bad too. We will not be able to hammer that steel into shape. Not too hot, not too cold, just so."

As they are talking the forgemen position the jib crane and prepare to remove the billet. With practiced ease the ponderous billet is lifted from the furnace, swung to the hammer and rested on the anvil in one continuous motion. Jan points with his gloved hand to where he wants Ben and then signals the hammerman that they are ready. *WHOMP*. The tup strikes the glowing billet. *WHOMP*. Again. *WHOMP*. The forgemen, jib crane operator *WHOMP* and hammerman work together, each anticipating the other, *WHOMP* without need for commands. *WHOMP*. Ben can feel every blow in his feet and arms. *WHOMP*. Steam hisses from the cylinder *WHOMP* atop the hammer with each return of the tup *WHOMP* *WHOMP*. With each blow sparks fly from the hot steel. *WHOMP*. Now Ben can see that *WHOMP* one end of the billet has been flattened and widened. *WHOMP* *WHOMP* *WHOMP*. The end begins to round out. *WHOMP*. Jan's arm shoots up in a signal to the hammerman. The tup ascends to the top of the hammer and stops. The billet-becoming-a-rod is swung back to the furnace.

Resting at the furnace, Ben realizes that he and the others are wringing wet, the sweat pouring from their bodies. Until now he's not been conscious of the heat. The dance of the forgemen took every bit of his attention and strength. Five of them had worked together as one. "How did you know when the billet had to be returned to the heat?" Ben asks Jan as they both blot the sweat pouring down their foreheads. "I can feel it in the steel," he says looking at his hands spread open before him. "I can feel it in how the steel moves under the hammer when is time to go back to the furnace." Jan can tell from Ben's expression that he hasn't been much help to the boy. "It is hard to know. But I tell you this. When you have been hammerman as long as I have, you will know too."

Work continues until the drive rod is done and, with it, the workday. Only Ben remains in the empty forge. From the distance comes the hiss of the showers, the banging of locker doors and the subdued talk of weary men. Ben crouches by the graceful rod laying on the forge floor. The heat it still holds from the furnace warms his face and hands. Nearby lay other rods for 1070. Just a few hours ago they were rough blocks of steel. The forgemen and hammerman's skill shows on every piece. Ben is amazed at the even surfaces; hardly a hammer mark is to be seen or felt. He tries to lift a rod to feel its weight in his shoulders, to complete his sense of it, but it won't budge. He laughs at himself, remembering that it has been worked down from a billet weighing nearly two tons.

The rods are scaly and ash gray from the repeated heating and hammering, but in Ben's imagination they have already become gleaming, silvery links in 1070's power chain. Through a haze of steam, he can see the piston rod and crosshead flying back and forth, and these rods stroking white-rimmed black drivers, and 1070 pedaling resolutely up a steep grade ahead of an endless string of cars.

The swirling clouds of smoke and steam are parting now, and with each day that passes Ben's image of his locomotive becomes clearer.

As Ben is about to leave, Jan calls him over to the hammer as if to show him something there. The tired, grimy hammerman pulls a delicate gold pocket watch from under his overalls and, to Ben's astonishment, lays it crystal up on the blackened anvil. Without another word, Jan grabs the operating levers. He raises the tup to its full height and brings it hurtling down toward the anvil. Ben closes his eyes to await the inevitable thud. But there isn't one. He waits a moment and then opens his eyes cautiously. Jan looks very serious as he motions Ben toward the anvil. "Now, you go look at what I have done." Ben walks over to the hammer, gets down on one knee and peers into the narrow space between the tup and the anvil. There is the thin silouette of Jan's pocket watch, the heavy tup hovering not an eighth of an inch above the gleaming crystal.

⬦ 6 ⬦

Machine Shop

BEN KNOWS the machine shop. For years he's been coming here to visit his father, and it's where—he decided some years ago—he'd most like to work. He can remember, as a little kid half the size he is now, wandering through the forest of gray and green machine tools. He'd arch his body way back and look up into the cobweb of roof trusses high above. An acre of roof floats up there on three-story-high glass walls of blinding white rectangles framed by slender columns.

When little, Ben always had certain stops he'd make on these shop adventures, and one he'd make without fail was "Pop" Stoltz's. Pop had a tall wooden cupboard next to his bench, more like a little store, really. Hanging inside the open doors were all sizes of caps and work gloves and suspenders. On one of the shelves was Red Man chewing tobacco, Pearson's peppermint snuff, matches and cigars in flat boxes emblazoned with colorful medallions. On another shelf, lower down, closer to Ben, were green apples, popcorn and caramel corn, Wrigley's P-K gum and candy bars. Ben was especially interested in these, of course, Hershey bars, "Damfinos," and his favorite, Planter's "Peanut Blocks."

Pop was a tubby, burly little man with but a fringe of gray hair around his glistening bald head. Pencils, rules, calipers, a corncob pipe, a scriber and a ball-peen hammer bristled from the pockets of his faded blue overalls, which seemed wider than he was tall. When Ben would appear,

Pop would always greet him with a "*wie geht's*," turn off his milling machine and get Ben to giggling with his Charlie Chaplin strut. They'd kid around for a time—Pop also did a great impersonation of slapstick comedian Ben Turpin, crossing his eyes, waggling his eyebrows and carrying on. Then he'd reach into his cupboard, as though he were doing a magic trick, and pull out a little ball of orange hard candy wrapped in cellophane, holding it up before Ben like a precious jewel (which Ben would later add to his collection in the small tin box on his dresser). With that, Pop would tousle the boy's hair, send him on his way and, hitching up his pants, turn again to his work.

Pop's not at his machine this morning. He retired a few years ago, but Ben sees him every spring at the company picnic in McBeth Park. It's early and the machine shop is just awakening. Ben's search for 1070's new castings and forgings leads him around a jumble of carts piled with steel stock, freshly cut bolts in neat stacks, bins of curly, iridescent blue metal chips and, everywhere, tools, all kinds of tools. In each machine is a piece of iron or steel becoming a finely finished locomotive part. The early morning sunlight flashes off a galaxy of polished, oily surfaces, edges and corners. On all but the darkest winter afternoons, the work bays are bright with the light needed for the most exacting work.

One by one, the men appear at their machines. They gather in twos and threes, sipping from cups of steaming

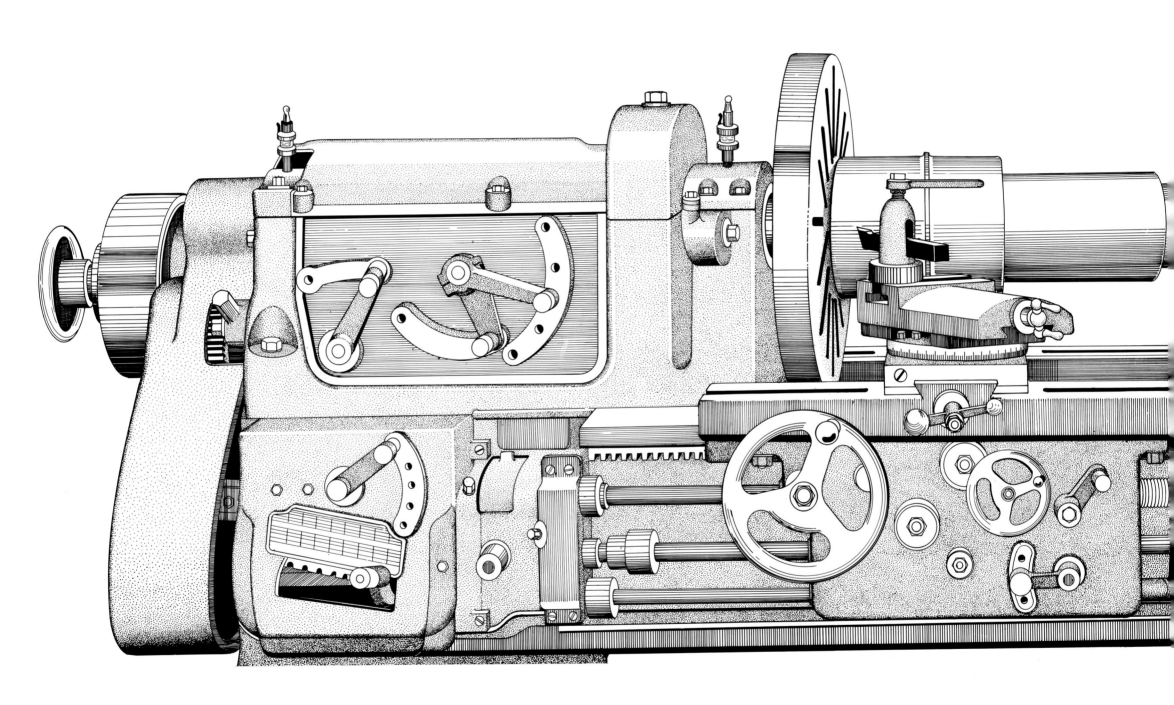

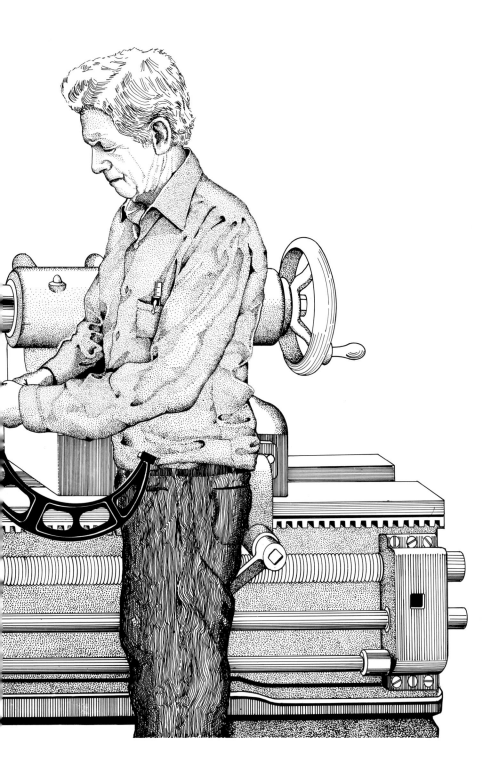

coffee and glasses of amber tea, speaking all the languages Ben has heard in Lima's streets and shops. Some are going through morning rituals of setting out tools and gauges in a row, polishing each lovingly with oily blue flannel cloths, setting up their work spaces. One man sings "Barnie Google with his goo-goo-googly eyes," as he chooses calipers and a scale from a shallow drawer in a dovetailed oaken chest. "Hunky" John is getting a head start on his lunch already this morning, the usual slice of white lard on a big round of onion between two thick pieces of coarse, dark bread.

Ben overhears snatches of conversation as he passes by the small groups. Talk is of good fishing holes and a favorite fly, Buster Keaton's picture show at the Majestic Theater, the latest antics of The Loco's softball team and, of course, the weather (it was 105 yesterday, and already in the nineties this morning). "The Babe's comin back" a deep voice proclaims around the well-chewed stub of a cigar. "Yep, he's on his way to another season like '21." (Ben, like most boys, knew the Babe hit 59 home runs that year.) "He's got 42 now; and he's sure not slowin down none." But the big news this morning is St. Louis first baseman "Sunny Jim" Bottomly, who batted in 12 runs in 12 times at bat yesterday.

Baseball is on everyone's mind these days. Al "Dutch" Miorin passes among the men at their machines collecting quarters for today's baseball pool (the Yankees and the White Sox). On the wall is an inning board filled in with the progress of yesterday's game. This afternoon the foreman will let one of the men call over to Ciminello's cigar store to get the latest scores—Mr. Ciminello is one of the very few people in Lima with a radio. A few times, when Ben's gone in for a soda pop, Mr. Ciminillo would take off his headset and let him listen to the faint, crackling voice coming magically over the air from far away Columbus.

Baskets have begun to appear at some of the machines filled with home-grown tomatoes, lettuce, radishes, onions, cucumbers, garlic and carrots, still spotted with black Ohio soil, for sale or barter. Warren Johnson always has a few jars of his special light wildflower honey at his bench. Ed Kozlowski and Leo Zanotti have brought some of their wine grapes, and Jan Stepek is just setting out a bushel basket of red-blushed pears.

A whir of electric motors announces the awakening of the overhead crane, hovering over the bay like a giant spider. One by one, the overhead line shafts come on and with each the *slap, slap, slap, slap, slap* of long leather belts between a pulley on the line shaft and each machine below. Everywhere, now, there is movement. Lathes turn, shapers stroke back and forth, drills spiral, planer tables glide to and fro, milling machine cutters bite into gray steel bathed in a milky stream of soapy water. Looking up toward the Department 12 loft, Ben can see the head and shoulders of his dad and Joe Park bent over their lathes. The shop is humming and another workday has begun.

An errand for Mr. Schnell takes Ben to a corner of the machine shop bay he's not explored before. There he's in for a surprise. The frame slotter is the largest machine he has ever seen. After annealing, locomotive frames are moved to the machine shop for the giant planer that will finish the faces and edges. Then they are moved here to the slotter, which finishes the inside surfaces of the slots that hold the drive wheel bearings, or drive boxes.

The drivers must be spaced exactly, so the spacing of the slots must be exact too. The two six-inch-thick frames have been stacked and clamped down on the table. This way, they will be machined together and be exactly alike. Each of the three large cutter heads is positioned over a slot. One of the machinists, Yannis Skouris, lowers the cutting tool to check the depth of the first cut. When Yannis starts the slotter, all three rams will cut on the

downstroke and then quickly return upward. Before the tool re-enters the work, the ram moves over so that each cut follows closely on the next.

Yannis beckons Ben closer to watch. Each downward stroke of the tool shaves off a slice of steel twelve inches long and three quarters of an inch thick as easily as the

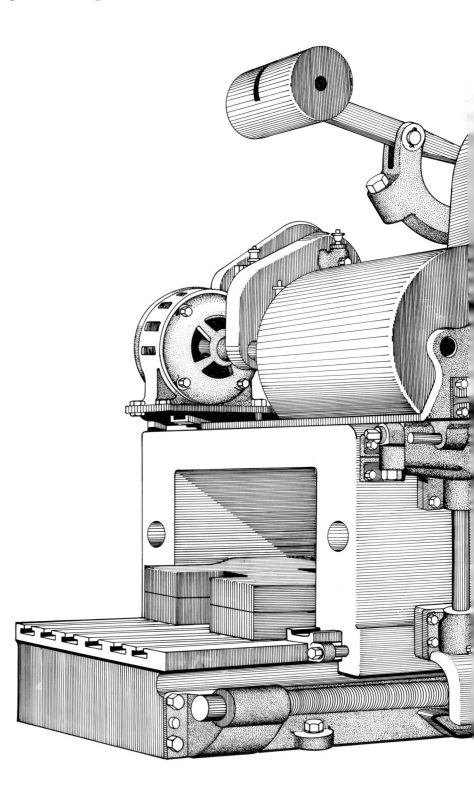

pattern maker's chisel slices through soft pine. Yannis digs down into his pocket, pulls out a nickel and, holding it up to catch Ben's eye, places it on edge on the massive iron slotter head. Again, the silver-blue cutters slice down through twelve inches of steel; even so the nickel does not move.

Ben walks the length of the slotter to watch the three heads working in unison. The side frames seem more massive than others he's seen around the shop. When he turns to the grease-smudged blueprints laying on the bench he knows why: there are the now familiar initials, "W.E.W." and Order No. 1070. Ben had always wondered what kind

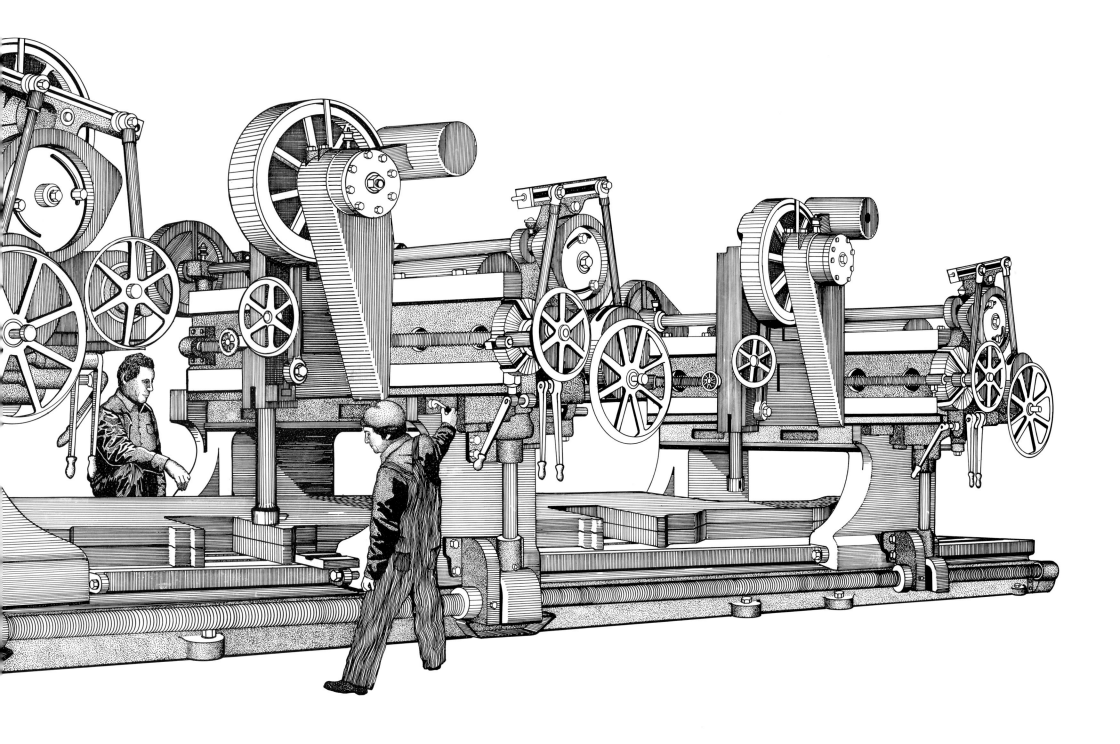

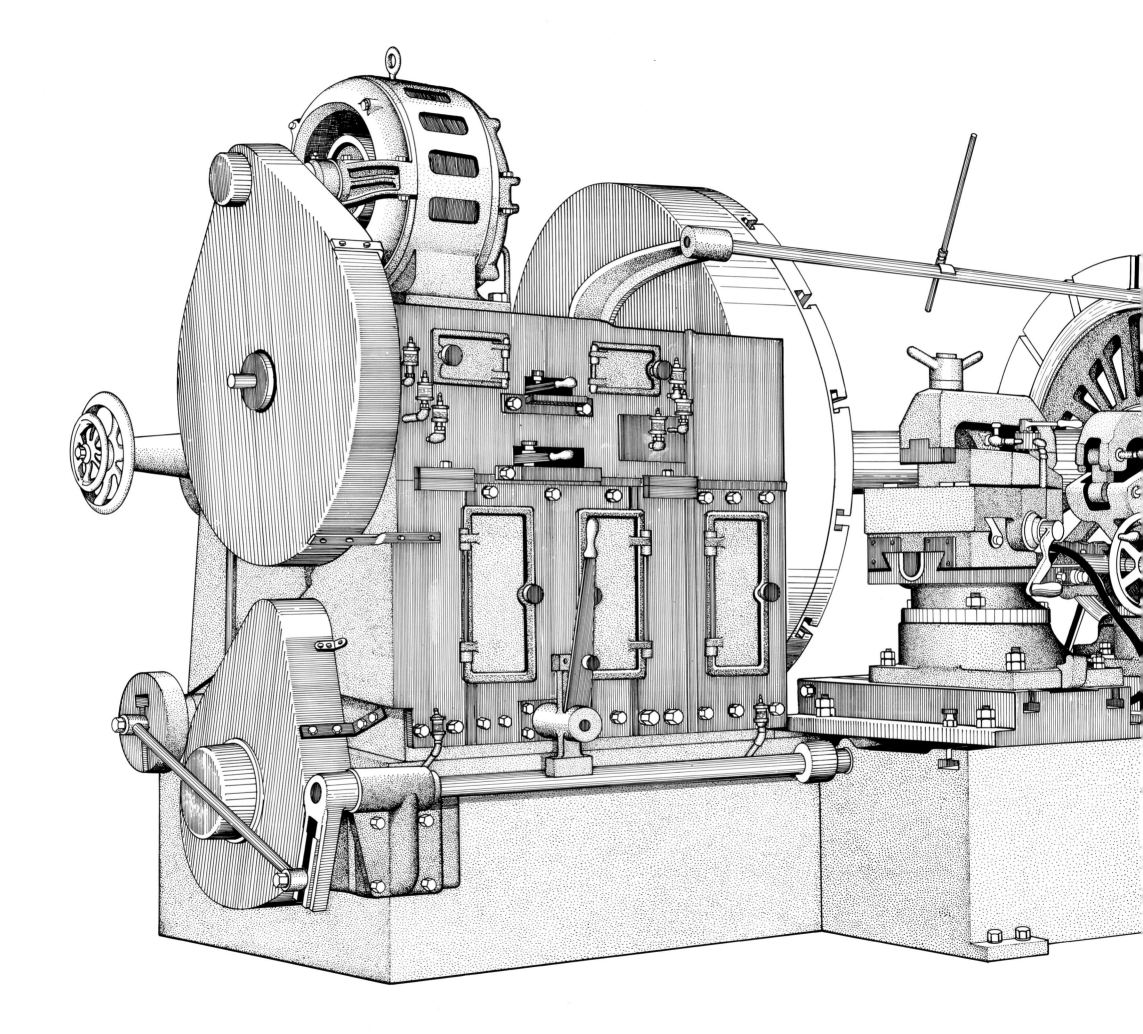

of machine could possibly make locomotive drive wheels—some as big as 90 inches in diameter—so perfectly round. Drivers not only need to be perfectly round, but all the drivers on the same locomotive must be exactly the same diameter. Any difference in the diameter of the drivers—even hundredths of an inch—will cause the side rods to bind and wear. Ben's wondering leads him to another corner of the machine shop and the wheel lathe.

A drive wheel is actually two parts. There's the cast steel wheel center, and the tire, a ring hammered from a steel ingot. The tire is flanged to keep the wheels on the rails. The wheel castings are first put on a boring machine that drills out the center hole. The hole is thousandths of an inch smaller than the diameter of the steel axle (about twelve inches). Shopmen call this a "press fit." The wheels and axle are put into a huge press. And the wheels are squeezed onto the axles under 200 tons of pressure.

Ben watches the next step, turning the rough castings round. The wheels on their axle are lowered between the two big, round face plates of the wheel lathe. The face plates are brought together until their pointed centers fit into matching holes in the ends of the axle. A "dog" attaches each wheel to a face plate so that when the face plates turn, the wheels and axle turn with them. Russ Park has set the tool rest so that both wheels will be turned to the same diameter. Skill is called for here; the huge wheel centers must be accurate to within thousandths of an inch.

No. 1070's eight drive-wheel centers, mounted on their axles, are lined up in the wheel shop awaiting their tires. Tires are forged from manganese steel, chosen for its very high tensile strength and wear resistance. Ben helps check the diameter of the tire with a caliper gage. And when he checks the inside diameter of the tire, he discovers that it's a few thousandths smaller. The tires are machined slightly smaller than the diameter of the center, just as the axle holes were. But instead of being pressed on, the tires will be mounted with a "shrink fit."

To get a shrink fit, the tire is heated with a circular gas torch in a spectacular ring of flame. As the tire heats it expands. At just the right temperature—when the tire turns dark blue—it is placed around the wheel center, tapped into place with large wooden mallets and left to cool. As the tire cools it contracts, and grips the wheel center so tightly the two become like a single piece.

The wheels and tires are returned to the wheel lathe. Russ will take the few thousandths of an inch off the tread that will make axle and drivers run perfectly true.

Will Woodard and Mr. Schnell have been suggesting to Ben some of the things he shouldn't miss. And among them is the work on 1070's new cylinders. Cylinder castings are the most complicated locomotive part of all. The intricate arrangement of chambers, passages and ports requires dozens of cores and a four-part pattern that fits together like one of those tricky wooden puzzles. Only a very few pattern makers and molders are skilled enough to make the molds for cylinders. Instead of being molded in a cope and drag, each cylinder half is cast in a deep concrete pit in the foundry floor. The casting will need 26,000 pounds of molten steel from five or six ladles all pouring at once.

As Ben watches, the machined cylinders are leveled with screw jacks on a steel surface plate six inches thick, embedded in the shop floor. The surface plate is absolutely flat so that all the machined surfaces can be checked. Finishing cylinder castings requires almost every kind of machine tool in the shop. Holes are drilled and tapped on a drill press. The cylinders and steam chests are bored true on a special boring machine. The saddle—the rounded part on top where the smokebox will sit—is shaped on a shaper. Even on these huge castings, the machining must be accurate to within $\frac{1}{64}$ of an inch (.016)—about the thickness of a printed i, l or t.

The helper's job is not as menial as it might seem at first. Each time Ben gets sent to the tool crib to pick up a tool or to the supply room for materials he also picks up a new word. What was yesterday a "thingumajig" or a "whatsit" he now recognizes as a stilson wrench, a broach, vernier caliper, reamer or chuck. "Hammer" could be a ball-peen hammer, a blacksmith's hammer, a beetle, a sledge hammer or a boilermaker's hammer. He learned the first day all about left-handed monkey wrenches. Once there were just "rivets," but now Ben can distinguish among cone head,

button head, steeple head and countersunk rivets. And because he has been looking at fireboxes, boilers and tenders every day (and sometimes crawling around inside of them), he is beginning to understand where each kind of rivet is used.

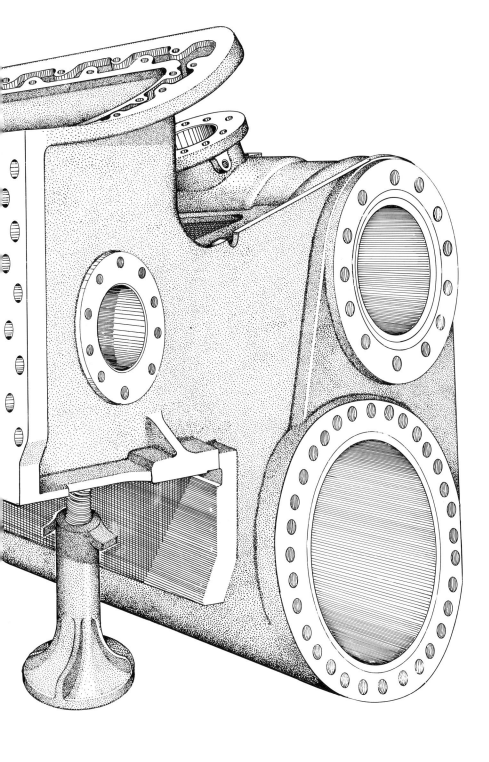

At The Loco Ben is learning by doing. He can hold an unfamiliar tool, or part or piece of material in his hand, feel its finish, heft and substance, and give it a name. He can watch it being worked. His instinct leads him to touch, to *feel* the difference between cast and wrought metal, brass and steel, surfaces finished on a planer or a grinder. He notices if holes have been reamed or if corners had been chamfered. He no longer describes surfaces as being simply smooth or rough. He can tell if a surface has been filed, scraped, rubbed, lapped, planished, fluted or knurled.

Everyone is Ben's teacher, like the machinist, Joe Park. Once he said to Joe that the hole he was drilling was "fatter" than the other. Joe smiled at Ben's term (he had used it himself as a kid and was corrected by his older brother) and then offered him another way of saying it— "bigger in diameter." Then Joe used the phrase pointedly in a couple of sentences so Ben would just pick it up.

For a youngster fascinated by machinery, The Locomotive is like a fabulous toy shop. Every time Ben discovers a new machine tool it becomes his favorite. And there are a lot of them—90 lathes, 70 boring machines, 45 planers, 42 milling machines, 105 drill presses, 15 grinders, 19 slotters, 20 shapers and an assortment of cylinder boring machines, saws, emery wheels and bolt and pipe threaders—each kind, in turn, attracting his attention.

All of this amuses Ben's dad. Twenty years ago, when Louis and his school friend Lars apprenticed together, they each went through the very same excitement of discovery. He ended up on the lathe and Lars on the milling machine where they are to this day and where Ben visits them frequently. Louis knows that Ben may not have a choice and that the foreman will decide to put him wherever he is needed. He trusts Mr. Schnell to make a good choice for his son, to consider his interests and abilities as well as the company's needs. After all, Mr. Schnell did it for him back when his apprenticeship was over.

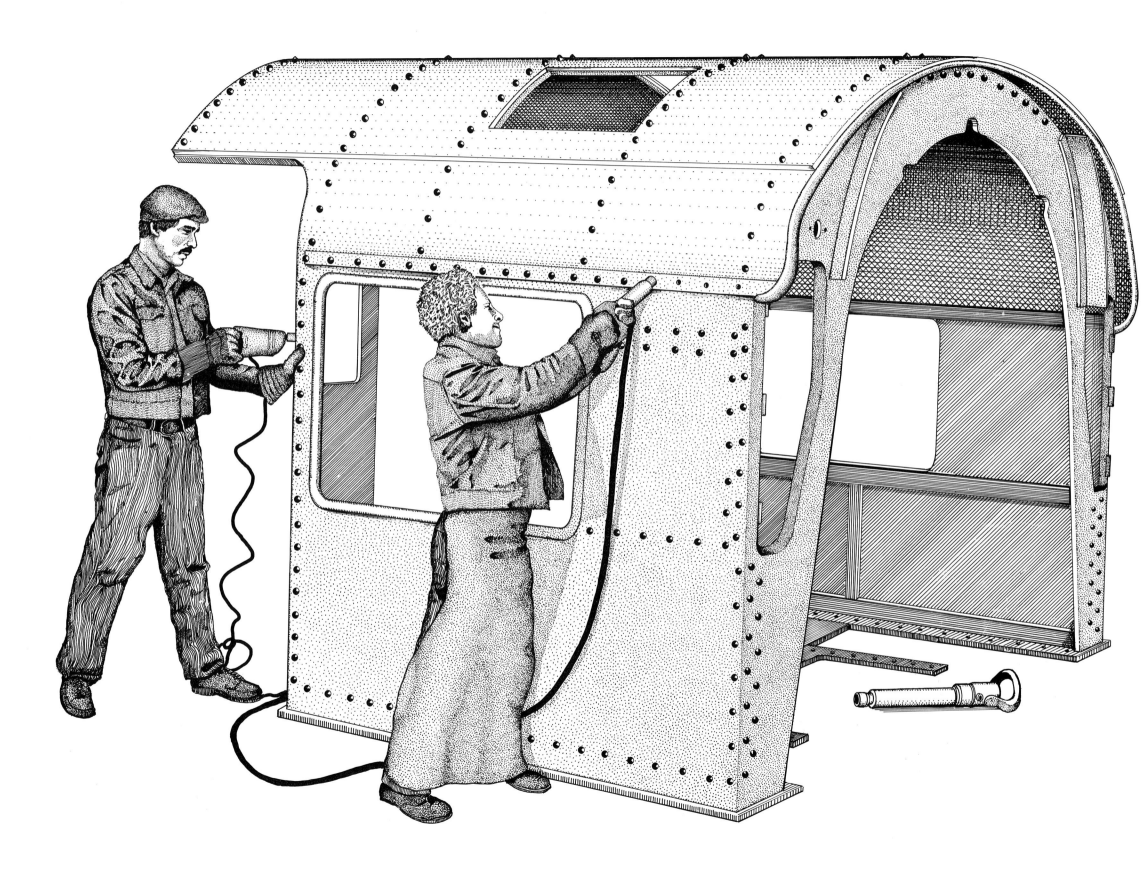

◇ 7 ◇

Boiler Shop

"You don't walk through the boiler shop," Ben has often heard Mr. Schnell say. "You walk over it." And he's found that to be no exaggeration. The busy, cluttered shop, strewn with odd metal shapes from some immense jigsaw puzzle, covers over two acres. It is the largest shop at The Locomotive and by far the noisiest.

The shop and everything in it seem bigger than life. Scattered about are dozens of steel tubes big enough to walk through. Each is a different length and studded with hundreds of rivets, staybolts and holes—locomotive boilers in the making. Great sheets of steel rolled into U-shapes mark the beginnings of fireboxes as big as rooms. Riveters lean from ladders and scaffolds high up on the boilers. Others walk along the backs of the steel monsters, the hoses for their pneumatic hammers trailing after them. Here and there showers of white sparks illuminate goggled welders cutting shapes out of flat plates or openings in boilers and smokeboxes.

The men who work here feel special, and others see them that way too. "Those boilermakers," Ben has heard his dad and grandfather say, "are a clan." They are all big men; their work requires that. But Ben detects something more—strong resemblances. Once, when he asked Mr. Schnell whether there was a chance he might work in the boiler shop, the foreman looked him up and down and answered without hesitation. "No, you're not built for it,

Ben. Besides, you've got no family there."

Ben is right about the family resemblances. Many of the men in the boiler shop are kin. It is usual for men in a shop to arrange jobs for their sons, or grandsons, or brothers or cousins. After all, that's how Ben got to The Locomotive. But it's different for boilermakers. Ben realizes this after meeting riveter Pete Rutaki. He is "Petro" in the shop because here he's among his countrymen. He is Petro, too, because he is among family. Actually, Ben was beginning to think that everyone in the shop was a Rutaki. He was about to change his mind when he first got to talking with Joe Seitchik. But Joe, it turned out, is Petro's brother-in-law.

Scattered around the shop are boilers of every shape and size. Some are steel gray; others have received their dull coat of red lead primer. All are covered with numbers, and arrows and instructions for the welders and riveters scrawled in white chalk. Ben recognizes among them the long boilers for the 2–8–8–2 malleys and one for an 0–4–0 switcher that could almost fit into the malley's firebox. He checks for the order number chalked on each one, and in a little game with himself he tries to guess how many holes have been drilled into a large firebox. He starts to count them, but soon gives up when he realizes they number in the hundreds.

A siren screams from overhead and Ben finds himself suddenly engulfed by a dark shadow. Startled, he looks up

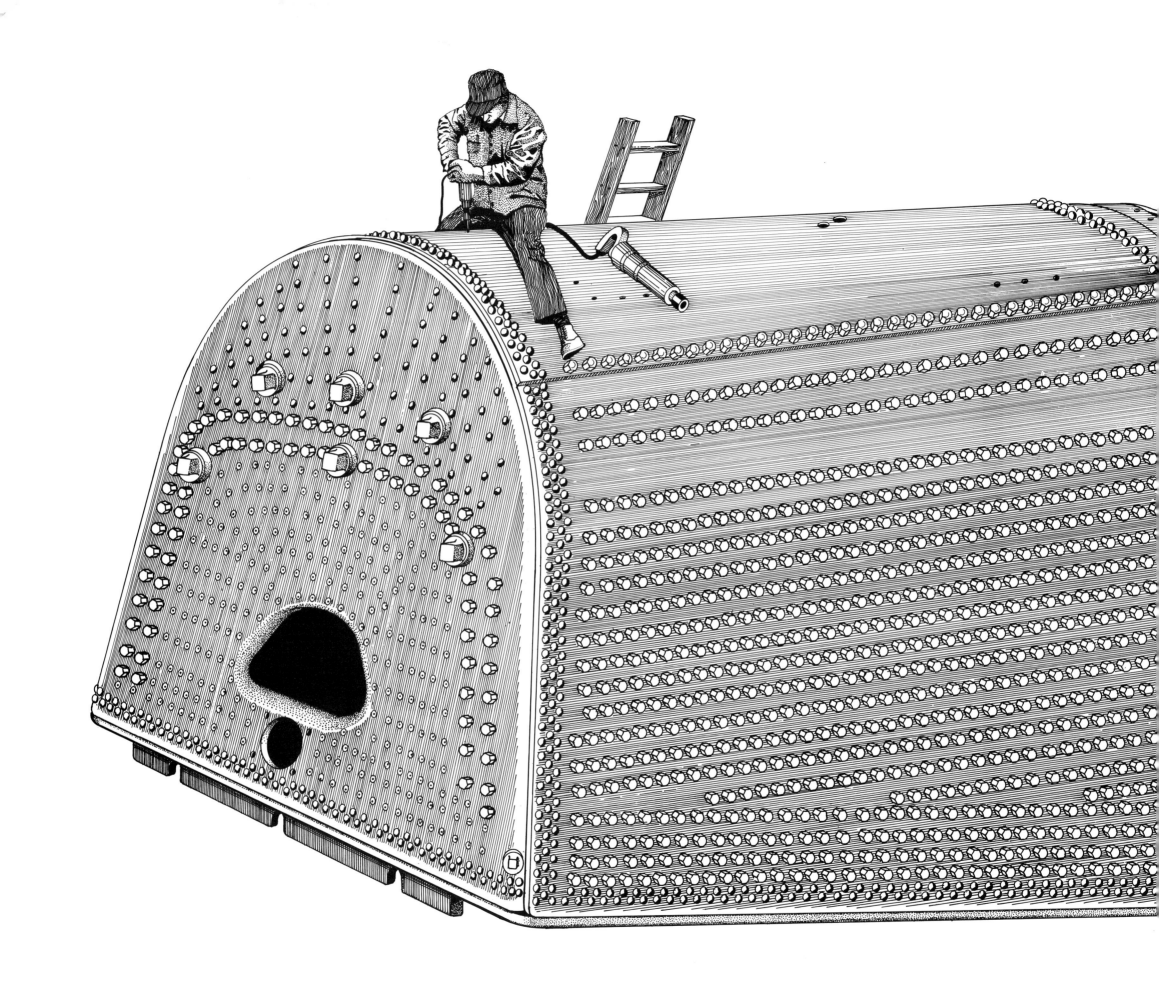

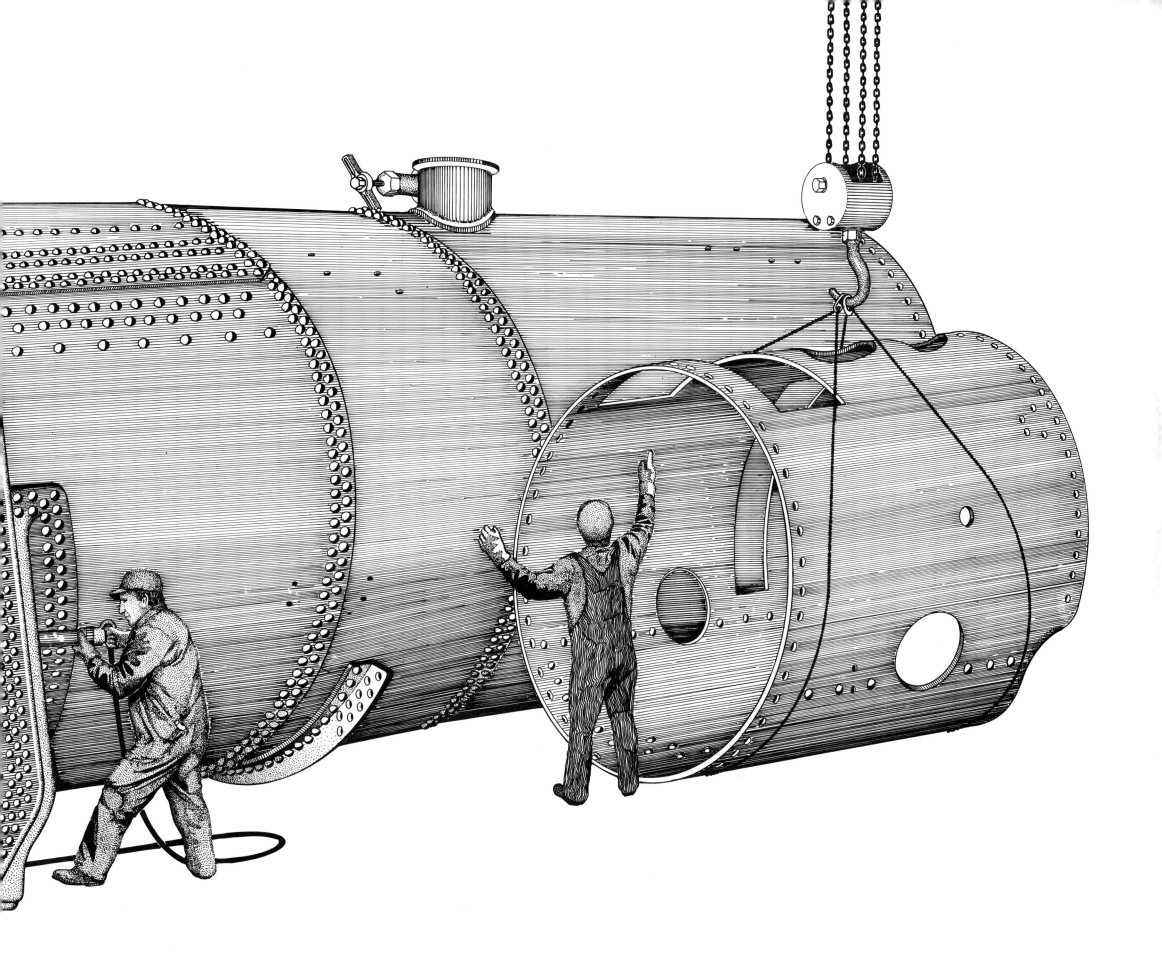

to see one of the long Mallet boilers hovering, swinging gently just above his head, suspended from the traveling crane. Then, with a whirring of electric motors and another howl of the siren, the boiler ascends and glides away to the far end of the bay.

Big tools are needed for such work. Huge "brakes" bend plate one inch thick to any angle. Powerful rolls shape sheets of steel twenty feet long and twelve feet wide into boiler courses and fireboxes. The racket is deafening. Presses stamp out domes from two-inch-thick plates as though they were Ben's little tinplate trains. Sledge hammers boom against boilers resounding like giant kettle drums. Metal crashes and thunders against metal. And through it all the *tat-tat-tat-tat* of dozens of rivet hammers. Ben realizes now why most of the older boilermakers he knows don't hear very well any more.

In his search for 1070's boiler, Ben clambers over stacks of steel sheets, through the tunnels of unfinished boilers and around heaps of stamped steel domes, flue pipe, smokebox covers and bolts.

"Heads up!" someone shouts. Ben ducks instinctively. An incandescent white comet flies over his head, then another. "Watch out now! You wouldn't want to catch one of these." Ben moves quickly over to a man at a small furnace. "This is a rivet heater," he says, placing mushroom-like button head rivets into the small furnace. He

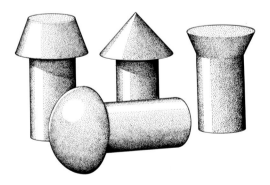

uses a pair of long handled tongs like the ones Ben has seen in his grandfather's forge. "Have to make sure these here rivets are just right. Have to be at white heat, ya know. Have to keep up with two rivet drivers."

The helper reaches into the furnace with his tongs and pulls out a glowing rivet. "Here goes." He lobs the rivet up to a worker on a scaffold next to a huge boiler who catches it in a small sheet metal box, like an outfielder holding his mitt under a pop fly. "He's a bucker. You can go on over to see what he's doin."

The boilermakers work together in five-man gangs—two drivers, two buckers and the man heating the rivets. After catching a rivet, the bucker picks it out of his box with tongs and places it into one of the rivet holes along a seam. Immediately, he presses a backup tool, called a dolly bar, onto the rivet and braces himself. From the other side a driver hammers the stem of the rivet into a head. In the old days rivets were headed by hand with a hammer. Now drivers use air hammers, or riveters, which can hammer lightly or with a powerful blow.

Ben's curiosity leads him to peer into every corner of the boiler shop, including the cavernous darkness of a firebox, where a surprise awaits him. A big hand at the end of a brawny arm reaches out from the firebox door opening and grabs him by the shoulder.

"Come on in here a minute," a gruff, southern voice echoes out from the darkness at the other end of the arm, "and give us a hand with this goldurned pipe." The voice sounds like it is coming from the depths of a big steel drum. And it is obviously at the end of its patience. So Ben climbs obediently into the firebox door opening. He feels a lot like Alice coming through the looking glass, for on the other side he encounters a bizarre scene.

In a long tunnel lit by harsh work lamps, two men struggle to wrestle a heavy pipe into place, casting monstrous shadows around the arching walls of the great boiler. The glaring light glances off the curved steel, setting each of hundreds of rivets in sharp relief. Every shout, every blow of a hammer pierces Ben's head as the three men lean, and bend, and push and pull against the long steam pipe in a grotesque, backlighted dance. An hour later the weary helper emerges grimier than he went in, knowing more about the inside of a locomotive boiler than before. He knows, too, that Mr. Schnell was right. Boilermaking is for bigger, stronger men than he.

Ben soon becomes used to the din of the boiler shop and is intrigued by the way the monstrous steel plates are fashioned into locomotive boilers. Before being formed into tubes, the flat sheets are drilled for hundreds of rivets, bolts and studs. Ben watches a welder cut holes in a sheet which will be rolled into a smokebox. At first the long, pointed blue flame seems only to splash off the steel plate. But soon a puddle of molten metal appears and, with a sputter, the flame penetrates the plate and shoots out below in a shower of sparks. Slowly, carefully the welder follows the chalk circle until he's back around to where he started. A steel disk clatters to the floor leaving a perfectly round opening for the smokestack. Two more cuts leave openings for the steam pipes to the cylinders. The flame slices through the one-and-a-half-inch steel plate and around the chalk line like a fiery knife. Another hole is cut for the pipes connecting the feedwater heater the locomotive will carry up in front of its stack. The welder turns down the cutting torch valves, extinguishing the hissing flame, and lifts his dark goggles up on his forehead, surprised to see Ben standing there. "Some show huh? Well, this ain't nothin. You should see it cut through two plates at once."

When Ben returns to check up on 1070's progress, he finds the boiler—firebox, dome course, taper course and first course—assembled. Petro has chained up the smokebox and signals the crane operator to lift it into position.

But there's a delay, and Petro looks up to see the crane operator shaking his head "no" and jabbing his finger up and down, pointing to the hook. Petro climbs up on the smoke box and sees the problem: one of the loops has not been secured tightly to the hook. He attaches the chain, climbs down to the floor and signals again. The operator signals thumbs up. To a whirr of motors the huge smokebox rises a few feet off the floor and swings over to the waiting boiler.

The courses of the boiler are put together with a shrink fit. The gang heats the end ring of the first course with torches and then telescopes the cold, smaller ring of the smokebox into it. Tapered steel drift pins are driven into several of the rivet holes to align the holes of the first course and smokebox. Then the torches are removed, letting the larger ring contract and tighten to a steam-tight seal. As soon as the boiler cools, the gang will remove the drift pins and ream out the drilled holes so that they are ready for the rivets.

Ben has watched enough riveting the past few days to understand what needs to be done, and he's found a new friend among the boilermakers. Petro points out to him the different kinds of seams boilermakers use. The courses are joined with a lap seam, made by overlapping the two edges. For the horizontal seams that are made when the rolled sheets are shaped into a cylinder, the riveter will make a butt welt seam. The butt welt seam, Petro explains, allows the boiler to expand when it gets hot and contract as it cools without pulling itself apart or distorting.

"So, is only one way to learn the riveting. I give to you a hammer. Here. You try. You can rivet." Ben handles the pneumatic hammer clumsily, frightened of it. It's heavy, yet in Petro's big hands it looks like a toy. Ben's not quite sure he wants to go through with this. "My boy he was plenty scare the first time he do the rivet. Ya, but you strong like him. Come, come. I give lots help for you."

Ben climbs up into the smokebox with Petro close at his heels, and then they pull a hammer in after them. The two stand under the daylight-bright rivet holes running from their feet up over their heads like a starry constellation. Behind them are the round openings to the hundreds of flue pipes coming from the firebox.

"Remember what I tell you. Hold the barrel of the air hammer in line with the rivet. Hammer not so hard at first until the head starts to form. Too hard at first the bucker can't hold on. Move the die around the rivet so it makes nice round head. But not so much that the die cuts into the sheet. Nice round head, huh? Hammer until the rivet is cool. You'll see, it turns red. Okay?

"So, comes hot rivet now," Petro says, making sure Ben is ready. Ben nods his readiness trying to keep in his mind all of Petro's instructions. "Okay, let's go. Rivet!" Presently, Ben hears the clank of the rivet in the metal box and, in another instant, a star white shank appears in one of the holes. With a reflex that surprises him, Ben reaches up with the heavy hammer, places the die squarely on the shank, leans into the gun and squeezes the trigger. *BLAM-BLAM-BLAM-BLAM*. He widens his stance, shifting his legs apart to keep from being thrown back. *BLAM-BLAM-BLAM-BLAM*. Outside, a helper slams the steel boiler again and again with a sledge hammer *BLAM-BLAM* forcing the seam together. *BLAM-BLAM-BLAM*. His ears roaring with the percussion of the hammering, his eyes fixed on the rivet, Ben feels like he is holding onto a wild, writhing, wriggling steel animal.

"Is good! Is good!" Petro shouts out. Ben releases the trigger, pulls the hammer back and sees that the yellow shank has become a round, dull red mushroom button. The driver is smiling. "Da, is good you rivet."

Ben learns more about riveting. He has a chance to try his hand with the dolly bar bucking for Petro. And he finds out why Petro and the other drivers are so careful with their work. Riveting does more than just hold the boiler courses and firebox together. The riveted seams must be tight against high pressure steam. The boiler will expand and contract with heating and cooling. And the steam pressure inside will reach 240 pounds per square inch.

Ben and Petro work together the rest of the afternoon. When the siren sounds the end of the workday, Ben realizes how dark it's become inside the smokebox shell. It's getting on toward winter now and the workday is getting shorter. Petro grabs him by the arm and, with a sweeping gesture, shows Ben that he's completed both courses of rivets around the smokebox ring.

"How you feel 'bout that, huh? How you feel 'bout the rivets?" Ben looks around him at the arch of rivets over his head and the two begin laughing. "Okay," Ben says, punching Petro on the shoulder, "I feel okay—but I don't think I'll be able to use my arms ever again."

"So you rivet this locomotive," Petro tells Ben as the two climb down out of the smokebox into the rapidly darkening shop, "and the 1070 boiler it is finish."

❖ 8 ❖

Beginnings

THE VAST, open shop is still too dark to begin work, still quiet at a morning hour which, during the longer summer days, would have been a hubbub. A winter-gray curtain clings to the towering walls of windows. Wind-driven snow cascades down the frosty panes and disappears into the white banks from last night. From above, flurries of flakes drift in through the clerestory windows and down to the icy-steel locomotives taking shape below.

With the growing brightness come hollow echoes of footfalls, banging doors and voices into the quiet. A figure appears bundled in scruffy, wooly sweaters. He drops wads of newspaper into a charred black drum, throws in a match, adds some kindling and then chunks of coal to the leaping flames. Ben's fire seems to draw other workers out of the frigid dark shadows of the shop. One by one, they join the circle of warmth, rubbing their frozen hands over the fire, stretching against the morning stiffness. Clouds of vapor rise from their shivering voices and the steaming cups of coffee clutched tightly between their palms.

Talk this morning turns to the deep snow they awoke to and scenes along the road of horses pulling Tin Lizzies and motor trucks out of the drifts. There's a reckoning of who got their firewood cut and stacked in time and who didn't. And, of course, there's talk of important happenings. It is duly noted that there's ice on the ponds and there'll be skating this weekend. And someone with a radio describes Red Grange's latest game in tones of disbelief. "Four touchdown runs—95, 67, 55 and 45 yards—the first four times he carried the ball!" It's 25 degrees outside and not much warmer inside. "It must be hard to work in here these winter days without any heat," Ben says to the venerable, gray-headed mechanic standing next to him. "Well, sonny, you know what the old-timers would say about that: 'If you worked as hard as you ought to, you wouldn't get cold.'"

Neither the worst blizzard nor the coldest cold could have kept Ben away from the erecting shop this morning. This is where locomotives are born. At every one of the thirty erecting pits, save one, stands a locomotive in some stage of construction. This is where the work of the entire plant—pattern shop, machine shop, foundry, smith and hammer shop, boiler shop, truck shop—comes together. It takes about seven days to assemble a locomotive. And almost every day a complete, new Mogul, Consolidation, Mike, Pacific, Mountain, Prairie—anything from a little 0–4–0 switcher to a mammoth 2–8–8–2 Mallet—steams out of the shop to the test tracks.

But Ben is not here to see any of these locomotives. He is standing expectantly, hands in pockets, between two empty rails set into the concrete floor. Nothing appears to be happening here, yet the air is charged with expectation. The empty tracks seem to hold out the promise of something to come, like seeds germinating unseen under the soil. Ben can feel it this morning, pacing back and forth, searching first one end of the track and then the other, waiting, looking, waiting.

"I was certain I would find you here this morning." Ben spins around to face Will Woodard looking bigger in his bulky overcoat, a roll of blueprints under his arm. It was Will who arranged to have Ben assigned to the erecting shop for the next few days. "And I'll bet you were the first one here, too." Ben smiles bashfully and nods. Shop foreman, Ray Schuck, joins them with an even bigger roll of prints. Then, like stage hands entering upon an empty stage from the wings, mechanics appear from everywhere, carrying heavy mallets, wrenches and prybars on their shoulders. Tools clatter and clang to the floor. Two men bring out carts filled with jackscrews, thick wooden blocks and timbers, bolts, doughnut-sized washers and nuts as big as Ben's fist. Another man wrestles a tangle of air hoses toward the erecting pit. A couple of sawhorses are dragged over and then a large board. The engineer and foreman unroll their blueprints and talk of changes and scheduling.

Over the past few weeks all the parts for Order No. 1070 have been brought together here in the erecting shop. From the iron foundry has come a smokestack, handrail columns, steam pipes, exhaust pipes, a bell stand, caps, clamps, steps, levers, brackets, valves and hundreds of other details.

From the steel foundry have come the huge main side frames, cylinders, trailer truck frame, furnace bearers, crossheads, spring hangers, a dome, drive wheel centers, three of these, five of those, twenty-two spring hangers, ten of this, twenty of that, on and on through a list pages long. Here, too, is the work of the forge, axles, cranks, side rods, main rods, crank pins, links, crosshead guides, piston rods all thickly coated with grease, 1,829 staybolts and a monkey wrench for the engineer's and fireman's toolbox.

Here and there in the black and gray jumble gleam polished brass and bronze jewels—oil cups, valves, petcocks, a builder's card, bearings, name plates and a bell weighing 83 pounds.

Some parts have been manufactured elsewhere and shipped here in wooden crates. The booster engine, grate shakers and fire doors have arrived from Franklin Railway Supply. Several crates are marked Elvin Mechanical Stoker Company. The Baker valve gear comes from the Pilliod Company, the headlight from Dressel Manufacturing Company, the marker lamps from Handlan-Buck and 1070's steam-powered generator from Pyle-National. The lead, trailing truck and tender wheels were shipped from Carnegie Steel and the couplers from National Malleable Castings. And somewhere there's a little box containing a bright brass whistle from the whistle makers, the Hammet Company. There are even parts from a friendly competitor over in Philadelphia: 1070's spring hangers are Part Nos. 892A63 and 892B63 from the Baldwin Locomotive Works.

Dominating everything here is the work of the boiler shop. Besides the boiler and firebox are domes, cover plates, runningboards and steps, brackets, a cab, tender tank, ashpan hoppers, cab ventilators, and the firing deck.

On and on go the lists. There are seats and cushions for the enginemen and a locker for their clothing, coils of electrical wire, a headlight switch, coat and hat hooks (four), a length of white whistle cord, flags (two each, red, green and white), a #4 scoop shovel, a coal pick, a broom. And over in the engineering department, on Will Woodard's drawing board, one special piece—a polished brass escutcheon, a five-pointed star in the Lima diamond, for the feedwater heater tank on 1070's forehead.

As each part arrived at the erecting shop, stenciled with the order and drawing number, it was checked off on a long list. Ben and other helpers were sent running from shop to shop—a little faster each day as the scheduled

erecting date grew closer—to hunt down missing items or arrange for delivery. Every piece is accounted for this morning, laid out in stacks and piles and wooden crates, in rows, like the pieces of a puzzle, awaiting the erecting gangs. In all, there are twenty-five thousand parts.

So the work can begin. The job of erecting 1070 will not be very different from assembling any other locomotive in the shop. Will Woodard's refinements leave the basic structure unchanged. The work goes on as it always has. The gang sets about positioning the jacks on the floor and bridging them with timbers.

A wailing siren announces the arrival of the overhead crane. Ben looks up to see one of 1070's long main side frames hanging above, swinging gently, machined and oiled surfaces shimmering in the winter light. A signal from gang leader Walter Gray brings the side frame down to within an inch of the trestles. The craneman waits as the crew swings the frame into position, and then sets it down with a solid thump. The empty hook ascends and the giant crane glides away. A few minutes later it returns with the second side frame which is positioned alongside the first.

The crane operator is a member of the team far below him on the erecting floor. He knows the sequence of assembly, which parts come when, where they are to be positioned. Although he relies on hand signals, he knows the work so well that he anticipates every move. One by one, the craneman brings the front deck casting (3,345 pounds), guide yoke crosstie (700 pounds), brake fulcrum crosstie (755 pounds), brace crosstie (500 pounds), reverse shaft crosstie (715 pounds) and the rear hinge castings (5,075 pounds). Each is bolted between the two side frames but loosely, so that all the pieces can be shifted and aligned.

Walter explains to Ben that the next step is to get the frame level. "You can help." He hands Ben the end of a long hose to which a water glass has been attached and

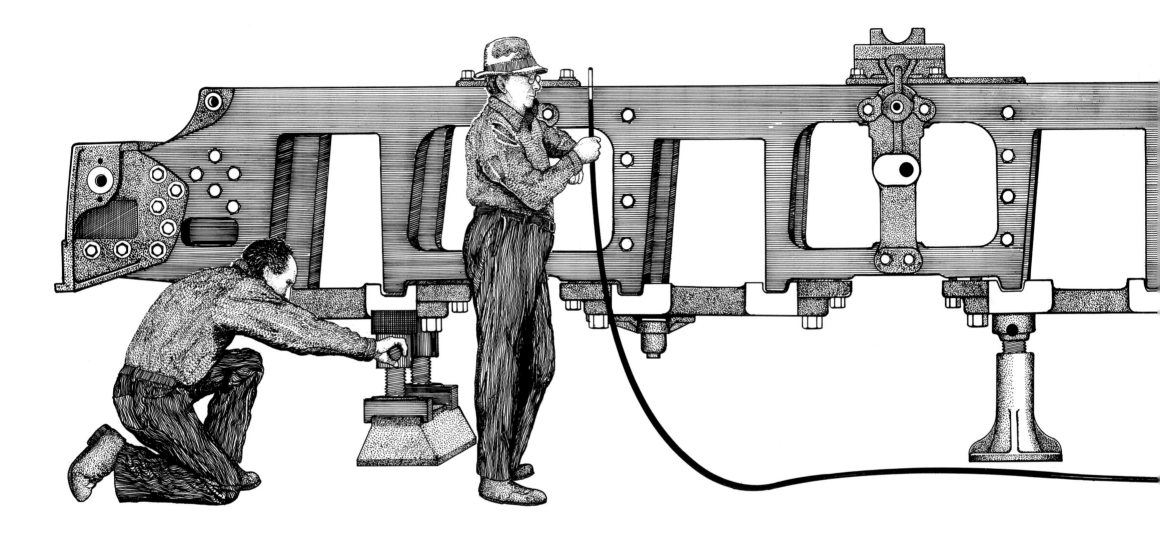

sends him to the front of the frame. "Now hold up the glass so that the water level is in line with the top of the frame. I'll do the same at the back end." The jack at the rear is adjusted until the top edge of the side frame is level with the waterline in both glasses. Then Walter sends Ben crawling under the level frame to the one on the other side. The hose now crosses under one side frame to the other. By raising and lowering the jacks on the side frame to match the waterline in the glass, both frames become level with each other.

No sooner are the side frames leveled than the overhead crane appears, this time with the enormous cylinder cast-

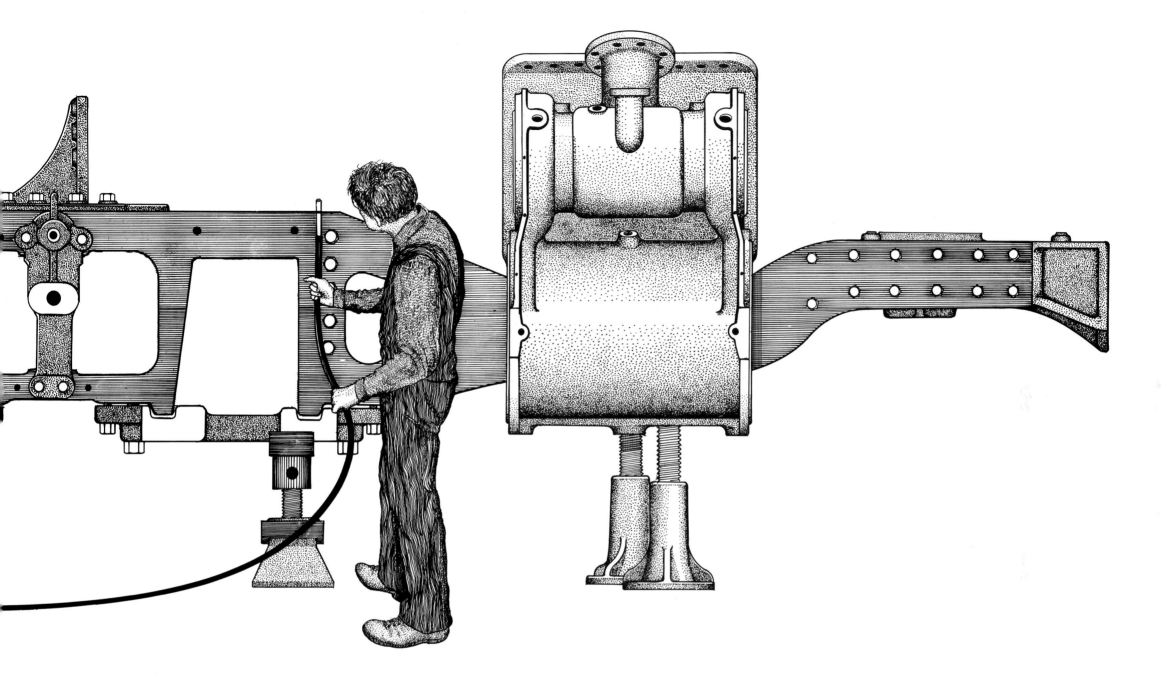

ing, weighing 13,235 pounds, swinging from its hook. The cylinders are slowly lowered as Walter, Ben and the rest of the gang guide them into position astride the main frame. The bolt holes are lined up, and with a thud the cylinders are set down. Ben helps tap the bolts into the holes and bolts the cylinders tightly to one frame, pulling mightily on a big open-end wrench. On the other side, bolts are just placed in the holes loosely.

The foundation of the locomotive—main side frames, cross ties and cylinders—is assembled. The main frame and cylinders must be precisely aligned if the drive wheels, side rods, main rods, crossheads and piston rods are to run true. To check alignment a line is stretched through the cylinder bore to a jack at the rear of the main frame. The line is carefully checked to make sure it runs along the exact centerline of the cylinder bore and is perfectly parallel to the main frame.

That done, Walter lays a large version of a draftsman's T-square across both side frames, bringing the head up close to the line. The head of the square, it turns out, is not parallel to the line and shows that the left frame is ahead of the right one. A special screw jack is put between the two side frames and slowly opened—forcing the frames into alignment—as Walter watches the head of the square and the line. When the head of the square and the line come perfectly parallel he knows everything is right.

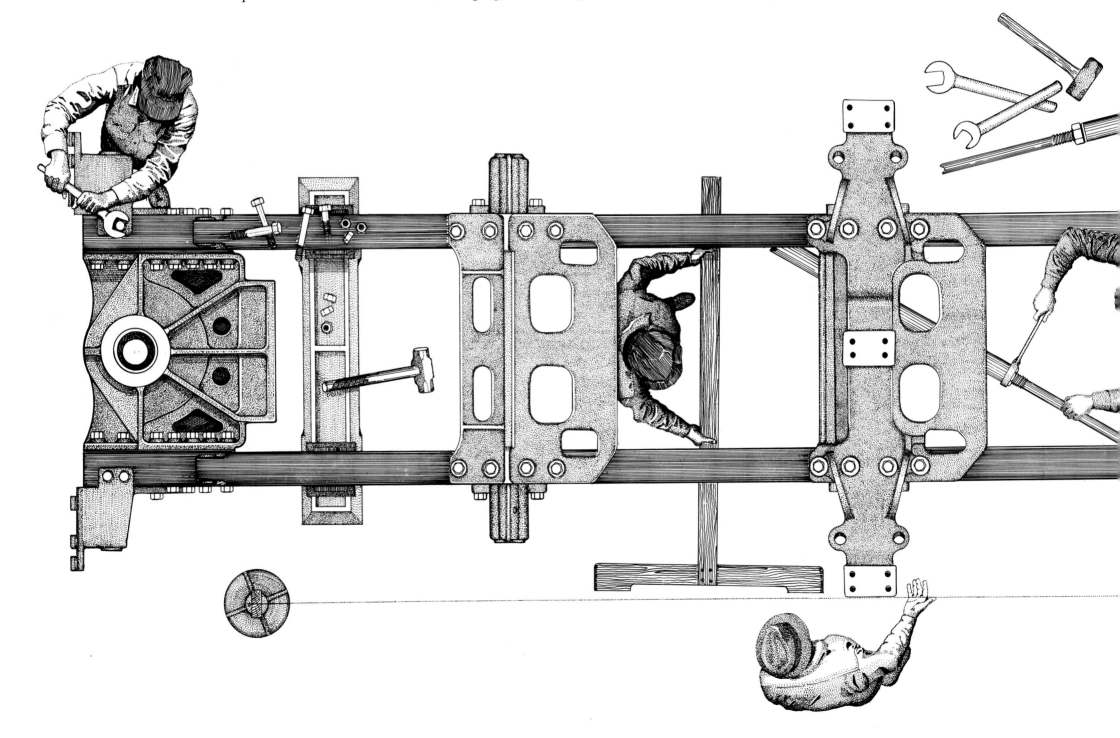

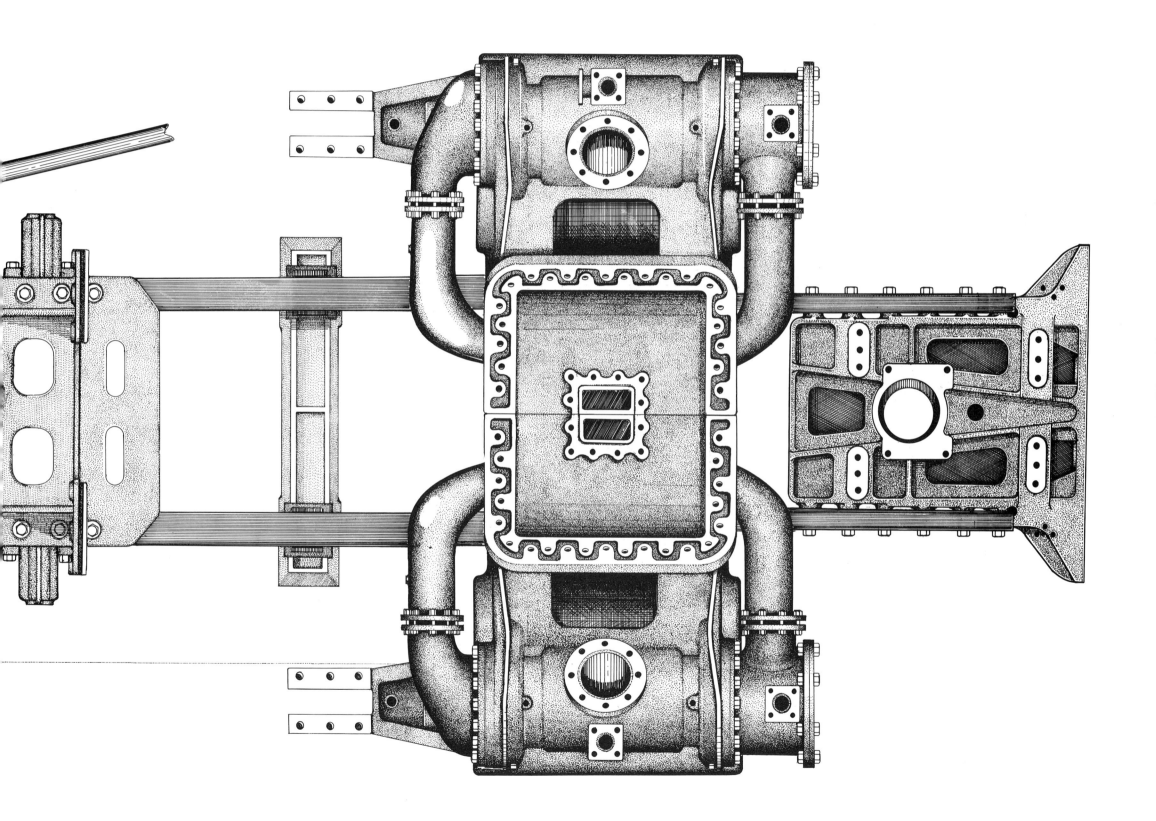

"That's it," Walter announces. "Let's get everything tightened up." The erecting gang swarms over the giant frame with wrenches, drills and tapered reamers. Because of the tremendous forces that vibrate and twist the frame, all of the parts must be bolted together tightly with no looseness or play. That means the bolts must fit the bolt holes exactly.

"In the machine shop," Ray Schuck explains to Ben, "all the holes are drilled undersized $\frac{1}{16}$ inch. If you look at the shank," he says picking up one of the heavy 2-inch diameter bolts, "you'll see it's tapered, bigger at the head, smaller at the threads. This tapered reamer will scrape out the hole so that it's tapered too, and makes a tight fit with the shank of the bolt. Here, I'll show you." Ray puts the sharply honed fluted reamer into the hole. Then, taking hold of the T-handle with both hands, he twists the reamer down into the hole, leaving fine steel shavings around the edge. Ray removes the reamer, blows away the chips and with a hammer taps the bolt down into the hole. "When this bolt is tightened it will really be tight, not just under the head, but along the whole length of the shank." Ben follows along behind, tightening down dozens of bolts, tugging at the long-handled wrench with both hands. As the bolts are all tightened, the side frames, cross ties and cylinders become a massive, rigid foundation for the new locomotive. And another workday comes to an end.

The gangs have left the floor. Alone in the silence of the erecting bay, Ben walks around the beginnings of Will's and his locomotive. He runs his hand along the machined side frames and coarse-grained castings, thinking about all that has happened today. In just a few hours an idea, Will Woodard's vision, has taken shape in steel. The shop has begun darkening to an early winter's eve, and for the first time since this morning Ben shivers with cold. He squirms back into his nubby sweaters, pulls on his wool cap and heads for the locker room. After washing up, he meets his dad at the machine shop and there is a lot to talk about as they wend their way home through the blue dark.

The work of the second and third days is not as lively. After the excitement of yesterday there are hundreds of details to take care of. The erecting gang breaks up into several smaller crews. One gang installs the two air brake cylinders between the side frames. These will be connected by a series of levers to the brake shoes on each driver. Another gang mounts the valve gear bracket on a cross tie, while two men assemble the power reverse shaft and levers. There's work to be done on the cylinders and, at the other end of the frame, the hinge casting to which the trailing truck will be pinned. Ben's jobs include holding things up, passing tools, running back and forth to the tool room, lifting parts and easing them into place, lending a hand and his strong shoulders to levering, pushing and tightening. He helps mount the strong leaf springs that support the drive boxes. They're a heavier version of the long, thin leaf springs he's accustomed to seeing between the axles and body of automobiles and big chain-drive coal trucks. Each spring weighs 376 pounds and is made up of steel leaves one-half-inch thick.

At the end of the third day, 1070's foundation is nearly complete and Ben has been working hard. He doesn't see anything remarkable about the way he's been working. It's the way he's worked since he was just a kid. But he and the other members of the gang are building a locomotive, a great and powerful machine, by hand.

When the overhead crane appears on the fourth morning it is with 1070's immense boiler and firebox—60,500 pounds—suspended on steel cables. There's some scurrying about as Ben and the gang drag blocks and timbers and jacks to the rear of the main frame. The front of the boiler, the smokebox, will rest on the cylinder saddle. But the rear, the firebox, will eventually rest on Will Woodard's special four-wheel trailing truck. In the meantime, the firebox will have to be supported by a heavy trestle.

The craneman lowers the boiler down to within a few inches of the cylinder saddle. Then, with ten pairs of hands guiding, pushing and pulling, with much gesticulating, grimacing and shouting, the boiler is set down. A gang climbs up into the smokebox with their tools and gets to work drilling holes for bolts that will hold the smokebox to the cylinder saddle. This is the only place where the boiler is bolted rigidly to the frame. The other connections between the boiler and frame must be flexible. These connections, called waist sheets, bend and buckle as the boiler gets hotter and cooler and expands and contracts. When hot the boiler will be almost an inch longer.

By lunch time, the boiler is mounted on the frame. Ben joins the rest of the gang sitting in a group near the stove. He opens his dinner pail, now battered with experience, to find his favorite lunch—two baloney and cheddar cheese sandwiches, some oatmeal cookies, and a quart bottle of milk topped with two inches of rich, yellow cream which he shakes vigorously. And an orange. Oranges in the winter—come to Ohio in the new orange refrigerator cars. Ben saw one for the first time only a few months ago. His dad took him to the icing station just built down at the freight yards. There they watched as workers with long, sharp poles prodded a parade of crystal blocks of ice along chutes to drop through the roof hatches at each end of the car and land with a resounding clunk inside. Ben puts the orange to his nose, inhales its fragrance and then returns it to his pail, a special treat for later.

Lunch is over. The men stand logy with sandwiches on hefty dark bread, cookies and cake, their overalls a bit tighter in the waist. They stretch and yawn and slowly amble over to their work. The siren, the familiar rumble, and once again the big crane is overhead. To the floor descends a two-wheel lead truck which is guided onto the rails. Four more times the crane returns, siren wailing, bringing pairs of drivers on axles to be lined up behind the lead truck. And a sixth time with the four-wheel trailing truck which is lowered to the end of the procession.

Now the crane sends down a second hook to the erecting gang. The chain hookers do their work carefully, step back and signal the craneman above. The boiler and attached frame rise up off the floor, more like a balloon than a hundred tons of steel, to hover just a few inches over Ben's head.

The work this afternoon seems to Ben more like play than any work he's ever done. The gang pushes together to roll the lead truck to the front of the engine, positioning it as close as they can tell to where it wants to go. Then, with a one, two three HEAVE, they lean into the first pair of drive wheels until 9,580 pounds of steel begins rolling down to the lead truck. Three more sets of drivers are rolled up one behind the other. Ben climbs up onto the spokes to unbalance the drivers just until they start rolling. Each set of drivers is carefully positioned under the slots in the frame into which the wheel boxes and axles will fit.

When it comes time to move the huge two-axle trailing truck into position under the firebox, another gang is called in to help. There are a few snide remarks about sissies who can't even move a "little" truck without help. But it takes no fewer than eight strapping mechanics huffing, grunting, snorting and gasping in three languages to get the truck with its booster engine—about 40,000 pounds—rolling toward its place behind the last pair of drivers and stopped.

Ben and the gang move the drivers and trucks into their precise positions under the boiler and frame dangling over their heads. Walter signals the craneman to lower the frame until it is just above the drive boxes. Again there's some positioning to do until everything looks just right. Walter signals once more, and with everyone watching carefully— HOLD IT! One of the drive boxes needs to be aligned

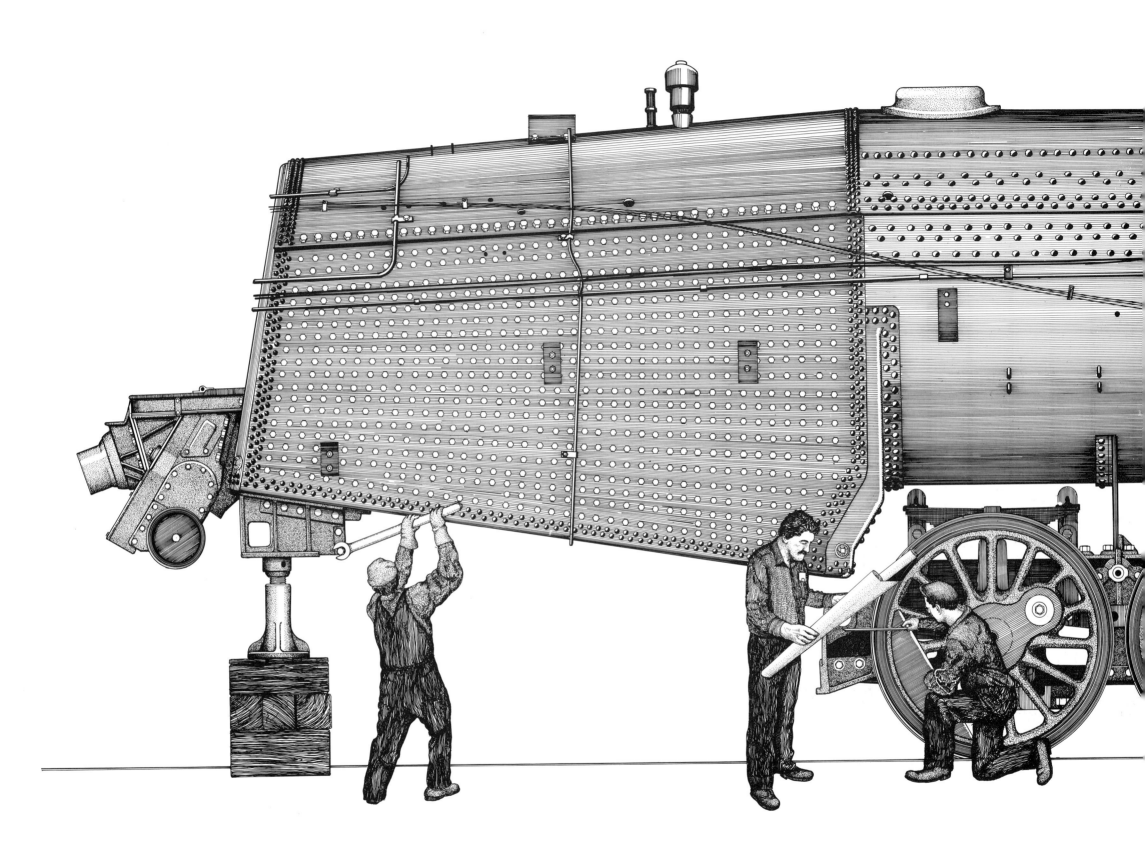

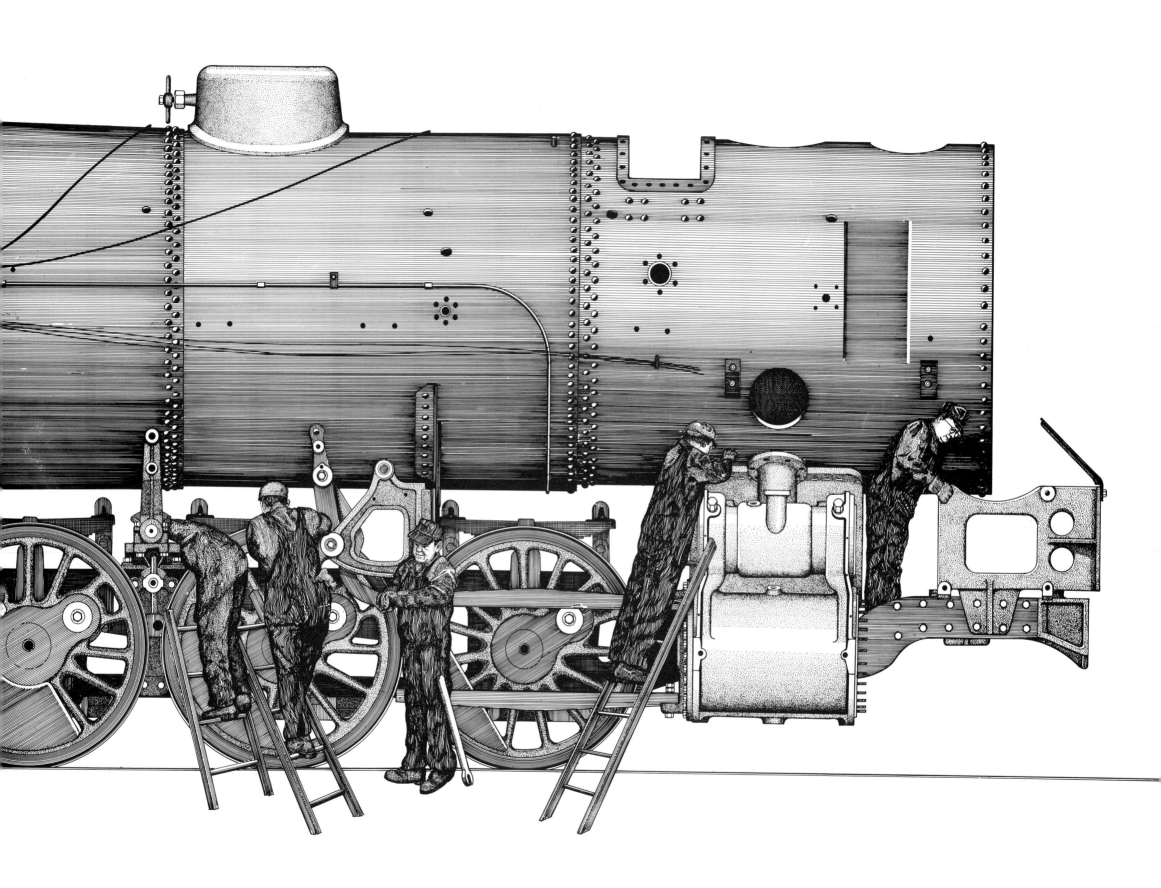

slightly. OK!—the frame inches down. Ben sees the big leaf springs begin to flex upward. Boiler and frame settle down and level out just as the cables go slack. No. 1070 is wheeled.

Day five. Ben doesn't feel like a newcomer anymore. Every day is still exciting—after all, this *is* the first locomotive he's ever built—it's just that it's no longer an event. The apprentice kid is an observer no more but a worker, a member of the gang. He knows how to do a lot of things without being shown how. And Mr. Gray just asks him to do this or that without even wondering if he can do it.

Two gangs converge on 1070 this morning. One works above on the boiler and firebox. There are the outside dry pipe connections to be made. The feedwater heater, stack, pop valves and generator have to be mounted along the top of the boiler. The throttle is installed and connected through rods and linkage to the back of the firebox where the cab will be. (It's in the paint shop and will be numbered and lettered today.) The runningboard brackets are bolted on, and the first of miles of piping is being strung all over the boiler. Ben's first job is to help mount the two brackets on the front deck that will hold a pair of air compressors for the air brakes. It's unusual; the practice has always been to mount the compressors on the side of the boiler. So, when the compressor shields are in place, they will give 1070 a unique or, as Ben says, swell front end.

The erecting gang is intrigued by this new locomotive, so there's lots of talk about it over lunch these days. While it goes together like any other locomotive, 1070's distinctiveness becomes clearer with each passing day. That huge four-wheel trailing truck with a booster could not go unnoticed. They knew the firebox and boiler were to be large—the guys over in the boiler shop have been talking about it for weeks— but until it descended from the overhead crane yesterday they hadn't realized just how large. In the new steel cylinders, the air compressors tucked in

under the smokebox, the outside dry pipe, in dozens of small details of design, Will Woodard's vision and skill are being demonstrated to these men who build locomotives.

Ben is learning a lot from these men. Every day there's something new to talk about with his dad. And he can feel in his hands easy movements that only a few days before seemed so awkward. But Ben is also learning things he's not even aware of. Work is one of them. He knows how to work, though he doesn't know *how* he knows. He's been learning about how to work since he was a kid. He's learned by doing. And he's learned from hearing adults talk about their work. Like what Mr. Schnell said to him one day last summer. He was telling Ben about the company's reputation among railroad men, that Lima builds the best locomotives in the world. When Ben wondered out loud how that happens, Mr. Schnell's answer was immediate and frank: "It's the workers, Ben. If you don't put out good work you don't work here anymore." And then Mr. Schnell explained why he felt that way. "When I was young, like you, Ben, my father worked on the railroad. He was an engineer." Mr. Schnell gestured toward the faded sepia photograph in a gilded frame which was always prominent on his desk. There was an engineer, looking a lot like Mr. Schnell does today, standing by his fireman on the pilot of an old diamond stack wood burner. "And I realized that if I did sloppy work, if I caused a wreck or a boiler explosion, I could hurt or kill my dad. So I put out good work." Ben thought of his Uncle John, and all of his friends' dads who worked on the railroad. There was no doubt in Ben's mind what was expected of him and, most important, what he should expect of himself.

When Ben returns with the gang from lunch, a surprise awaits them. Hanging from the overhead crane and descending to the rear of the firebox is 1070's new cab, freshly painted glossy black. On the sides under the window is A-1 in lettered lustrous silver. A-1. Ben likes the sound of it and repeats it to himself several times. "Hey, you guys, let's go," Walter calls out to his rubbernecking gang, "we've got a locomotive to build."

When the overhead crane appears again, it is with old friends. Touching down on the erecting floor right at his feet are the main rods Ben watched being hammered out at the forge. The chain hooker releases them and sends the crane on its way to return a few minutes later with a pair of side rods. While work continues this afternoon on the smokebox, boiler, firebox and cab, Ben's gang will assemble the main and side rods and the valve gear.

The splendid surfaces of the steel rods and brass bearings reflect the forgeman and machinists' finest work. Ben's first job is to wipe each crank pin clean of its gooey gray-green coating of protective grease. When he polishes off the last greasy smudges, a fun-house reflection of his face appears in the mirror-bright steel.

Ben helps guide the rods onto the crank pins. He can feel the precision as the rod bearings—or "brasses," as he learns to call them—slide onto the pins with but thousandths of an inch between them. When the rods are on, Ben puts a big washer over the threaded end of each crank pin, spins on a heavy nut and tightens it with a wrench almost as long as he is. Finally, he slips a cotter pin through a slot in the nut, through the crank pin and out the other side, and bends it around to keep the nut from coming loose.

Work continues apace this afternoon. While one gang has been assembling the side rods, another has been working at the cylinders. The crosshead guides have been bolted to the front of the cylinders. And the crossheads are now in their guides. Helping with the assembly and looking into the open cylinders, Ben is really able to see for the first time, to feel in his hands, the workings of steam power.

The piston and piston rods are in place. Ben imagines steam entering first one end and then the other end of the cylinder, the piston rod pushing back and forth, back and forth. Springy steel rings around the piston form a tight seal between it and the walls of the cylinder so that steam can't leak past. Since all the force of the steam acts against the piston rod, the rod must be very strong. It was forged from a billet of steel and then carefully heat-treated to relax all of the internal stresses. Ben watches as the rod is fitted into the crosshead and a key driven in to hold the two together.

The piston, piston rod and crosshead all slide back and forth together. When one end of the main rod is attached to the crosshead with a wristpin and the other slipped onto the main crankpin, A-1's power train is complete. The pressure of steam pushes the pistons, piston rods, crossheads and main rods outward. The force against the main crankpins turns the drivers.

Ben has been too busy with details to notice what has been happening with his locomotive. By the time the main cranks are in place and the eccentric straps connect them to the valve gear, the glint is gone from the polished rods. Now, they reflect the darkening gray of a late winter afternoon. When Ben finally steps back to have a look, to stretch and arch his weary back, he is overwhelmed. Before him stands a nearly completed locomotive on rails that, just five days ago, were empty.

Ben spends his weekend down on the frozen Miami & Erie Canal, clearing snow off the ice and skating the hours away. The fierce cold sends him often to the warmth of the bonfire crackling in the snow, and the laughter of friends. They talk of the simple delights of Christmas, winter and sweet things to eat. The fire and circle of friends is a little world into which the cold cannot intrude. The drifts have gotten so deep overnight that no automobiles are out. The quiet reminds Ben of his childhood when the only sounds of the plains winter were the wind, a dog barking now and then and the far-away, hoarse whistle of a train.

Monday, the sixth day. A hundred tasks await the gang. Up front, the headlight and marker lamps are mounted and connected. The overhead crane assists in mounting the two heavy air compressors on the front deck. Then they must be connected to the air reservoirs, brake cylinders and train lines with a perplexity of pipe. The shimmering, golden bell is lifted to its waiting hanger and connected to the automatic bell ringer. To the front deck the pilot beam and steps are attached. Up on the boiler the sand dome and sander valves are assembled and sander pipes run down the boiler to within a few inches above the rail in front of the first pair of drivers and behind the last pair. Cold-water pump, injector, check valve and feed-water heater are installed in a web of water pipes and steam lines. Much of the boiler piping was completed earlier, and the lagging of the boiler with blocks of magnesia insulation is just finishing up. A gang has already begun putting on the sheet steel jacket that smooths over the rivet and bolt studded boiler. Another installs the hose that will connect the locomotive to its water supply in the tender.

Ben climbs up into the cab for his first visit. The work here is almost finished. A profusion of gages, valves, and piping covers the firebox backhead. A mechanic remains to make last minute adjustments on the stoker and fire door. As Ben helps, the engineer's and fireman's seats are hoisted up into the cab and bolted in place. Completing his inspection of the work in the cab, Walter Gray pulls a soft flannel rag from his pocket and lovingly polishes fingerprints off the brass bezels and glass of the gages.

Then, suddenly, they leave and Ben is alone in the cab. He examines the cluster of gages in front of the engineer's position, grips the air brake control handle and then the power reverse wheel. He touches each of the controls, one by one, reading their functions etched into brass name plates. He runs his hand over the soft leather seat and, looking around to make sure no one is watching, slides up into the engineer's seat. Resting his elbow on the window sill, Ben leans way out, the way he's seen enginemen do, and looks ahead. He reaches out with his left hand until it falls comfortably on the cold, smooth brass handle of the throttle.

The erecting shop, the jumble of tools, locomotive parts, piping, bolts and jacks fades and resolves into a snow-blanketed Ohio countryside. The glare of the low winter sun flashes painfully white into the engineer's eyes. Telegraph poles, mile posts and familiar little towns along the Sandusky Division fly past the open window. The sounds of the wheels are muffled in the snow-covered roadbed. Ahead, two silver rails converge in the glittering whiteness. Ben looks down the long black boiler and ahead for block signals and grade crossings. Bitter cold wind stings his face and numbs his bright red ears filled with the awesome thunder of his locomotive. In the swaying, rolling cab rhythms of exhaust and clicking wheels are taken up by a wailful harmonica and driving banjo and Ben begins humming a favorite tune . . .

> Oh, listen to the jingle,
> the rumble and the roar,
> As she glides along the woodlands,
> through the hills and by the shore.
> Hear the mighty rushing engine,
> hear the lonesome hobo's call.
> As they ride the rods and break-beams
> On the *Wabash Cannonball*.

Fostoria junction is ahead. The semaphore blade stands straight up against the deep blue sky. The engineer reaches for his whistle cord . . . . "Hey, Ben. Ben? It's time to go home." The engineman looks down to see his father standing below on the dark erecting shop floor. Everyone has left the shop. It is quiet except for the wind in the clerestory windows above. "You didn't show up at the usual time, so I came over to see what's keepin you." Still somewhere in that distant scene, Ben climbs down from the cab, not on the wooden scaffolding, but on the newly installed steps. He takes his dad around the A-1, shows him the work he's done, explains what he's learned about assembling side and main rods. When it becomes too dark to see anything, Ben finds his lunch pail and jacket and heads out into the storm with his dad toward home, the warmth of the fire and supper.

Tuesday morning. Ben is at the shop early starting the fire in the steel drum stove. He just gets it going when he hears a commotion down at the far end of the erecting pit. Voices outside and the rattle of a heavy latch. The two tall wooden doors shake and then swing out and open. There on the track, framed in a rectangle of winter-clear light, is a massive black tender. It's for the A-1. Hidden behind it is the little 0–4–0 Ben has seen pushing and pulling around the shops. The switch engine's puffing and chuffing become more insistent and determined until the tender begins to roll. Only now can Ben see the little engine, peeking out from behind, dwarfed by the tender. With a grating, screeching of brakes the switcher slows and stops the tender within a few feet of A-1's cab.

No. 1070's tender is nearly forty feet long. It sits atop two six-wheel trucks. In its forward bunker is space for 18 tons of coal and its tank will hold 15,000 gallons of water. The erecting gang immediately prepares to couple locomotive and tender. Connecting them will be a stout hammered iron drawbar (530 pounds) and two safety bars (580 pounds each). With the help of the switcher, the tender is inched up close to the cab, the draw-bar and safety bars set into their pockets, and secured with forged steel pins. In another hour or so water, air and electrical lines and the stoker conveyor are connected. Emerging from between the locomotive and tender, Ben runs into Mr. Gray who's been looking for him. "Hey Ben. Joe Park's needing

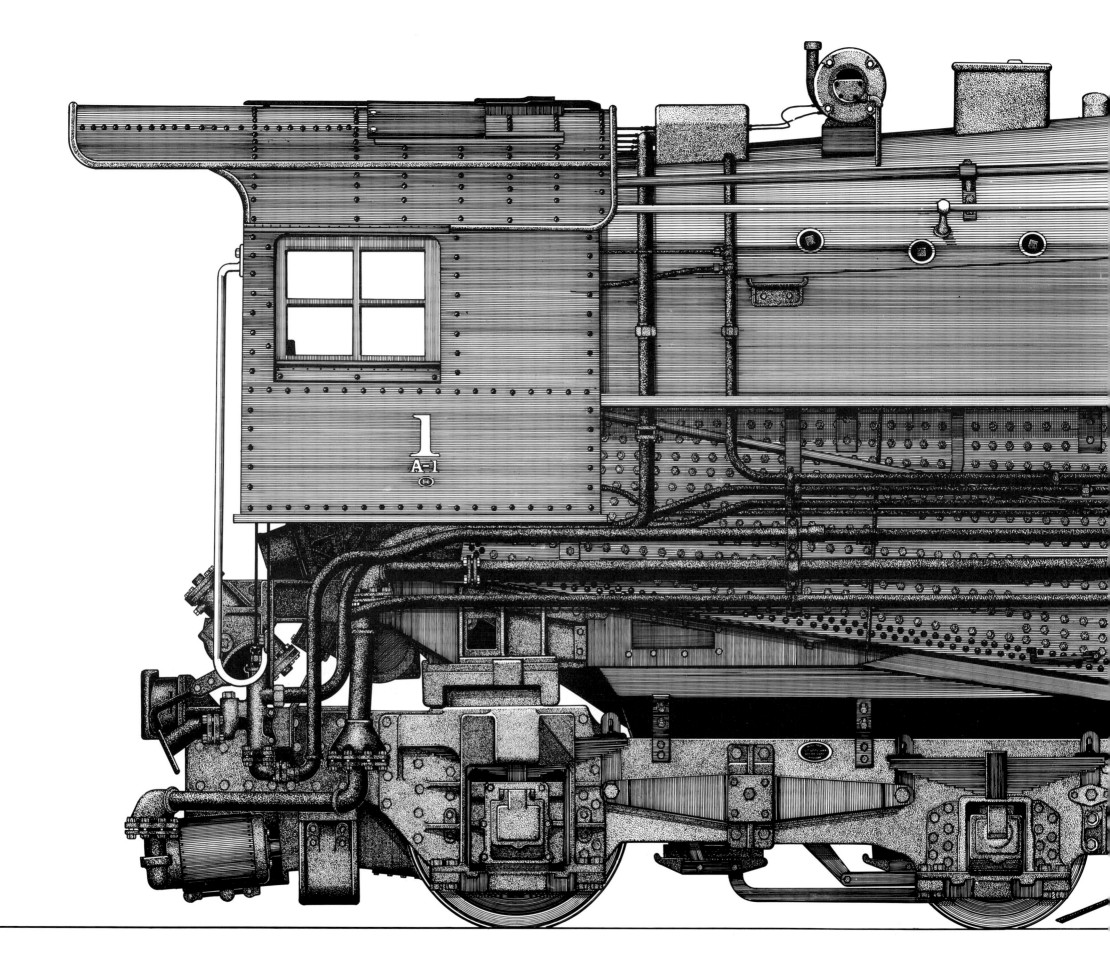

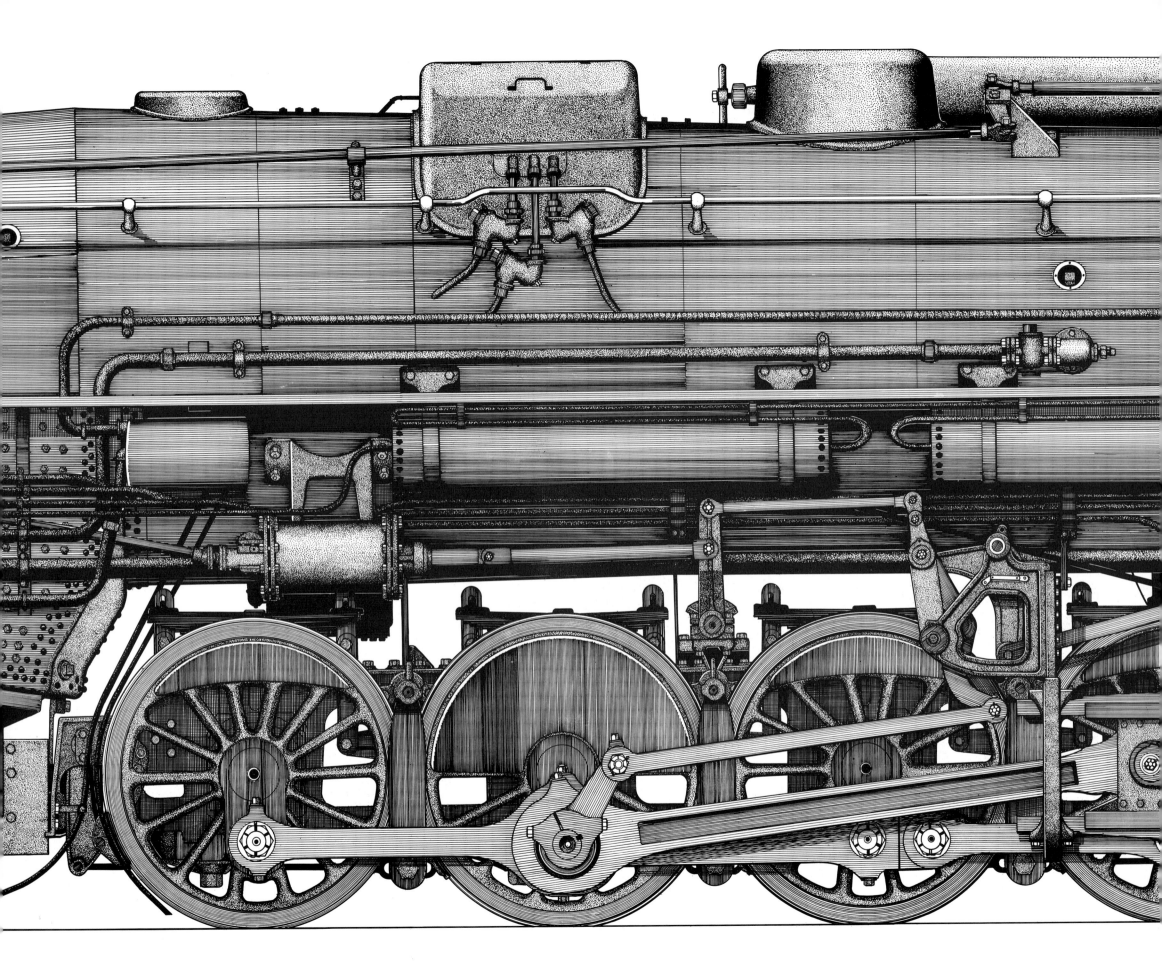

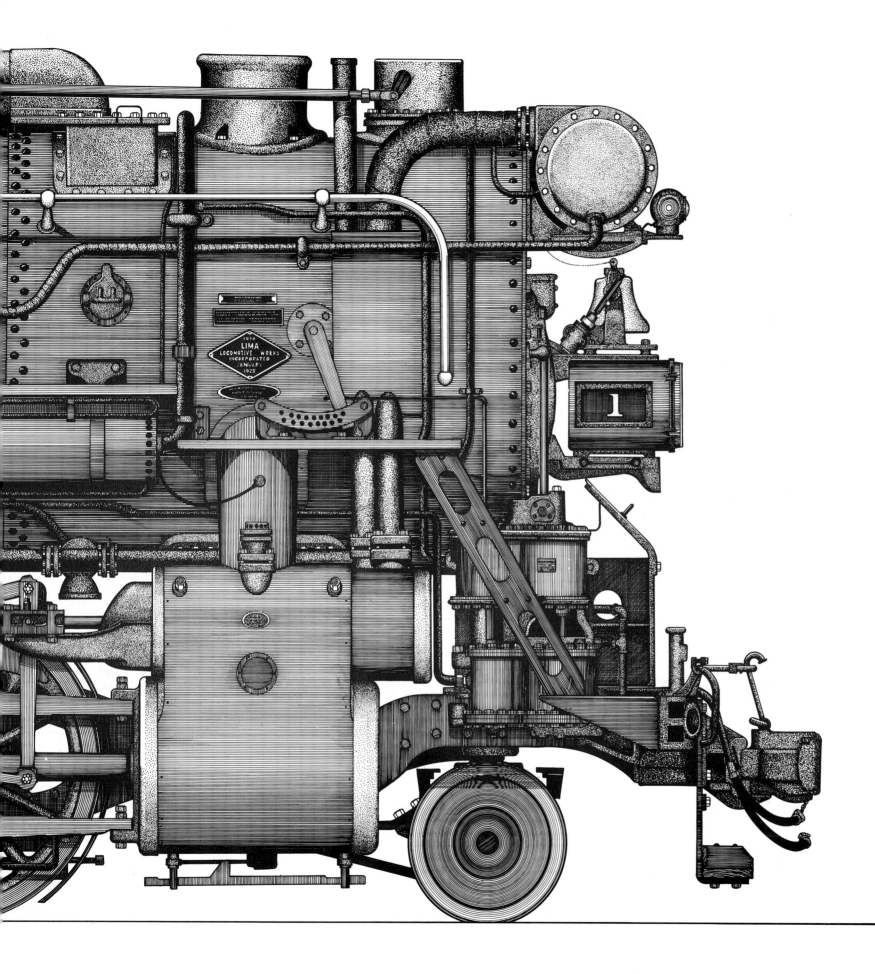

you in the machine shop. Better head over there now." Walter sees the boy's hesitance to leave. "Don't worry. You'll be back here in plenty of time to help us finish up."

◇

Joe explains to Ben that he has a crank pin he has to get out this afternoon. "Why don't you get the pin ready for centering in the lathe," Joe suggests. So Ben gets the square with the centering head from Joe's toolbox and marks the centers at both ends of the pin. Then he finds a center punch, places the point carefully at the intersection of the cross he's made and strikes it sharply with a hammer. Together they lug the heavy pin over to the drill press. Ben chooses the drill he needs, tightens it in the chuck and drills a small tapered hole at the precise center of each end.

There is something different about working with Joe today. Ben has helped his father's friend many times, but this time it's more as a machinist than just a helper. To Ben's surprise, when it comes time to center the pin in the lathe between the face plate and tailstock, Joe suggests that he do it. Joe asks Ben which tool he'd use. He approves the boy's choice and then lets him clamp it in the tool holder, checking to make sure it's angled just right and below the centerline of the work. Ben is surprised again when Joe gestures for him to take the first cut.

It is clear to Ben now that today Joe expects him to do the job. The master machinist watches as Ben carefully advances the tool into the spinning pin until the first blue hot curled chips spiral back from the point. Ben confidently locks the carriage feed lever and the tool begins moving slowly along the pin, peeling off the scaly dull surface.

When it comes time for the finish cut Joe takes over. As Ben watches, the tool travels slowly along the pin leaving behind it a glass-smooth cylindrical mirror reflecting the intent faces of machinist and helper. While they work, Joe asks Ben several questions about what he's been doing the past few weeks. He asks the boy to read the dimensions and tolerances to him off the blueprint. Then checks them for himself. All the time Ben is thinking that their conversation seems a little more pointed, more deliberate than in the past. Joe is his usual patient, good-natured self, but more serious somehow.

There. The pin is finished. Ben rubs his hand admiringly over the perfectly turned steel. Joe takes a pencil stub from behind his ear, licks the point and carefully letters a note on a piece of scrap paper. He reads it over, looking even more serious now, folds it and hands it to Ben. "Here you go. Take this to Mr. Schnell later this afternoon, when you're through in the erecting shop."

Ben returns to a changed scene at the erecting pit. A few mechanics work here and there making adjustments and installing the last few small parts. The rest of the gang is cleaning up around the pit. Ben is enlisted to help get wrenches and rivet guns, jacks, extra nuts and bolts and rivets, hammers and mauls and screwdrivers back to the tool room. Mr. Woodard has arrived along with engineers, vice presidents and other Lima officials.

The switcher returns and, inching its pilot closer to the rear of the tender, couples up. The doors at the other end of the erecting pit are opened. Ben imagines the little engine leaning into the huge tender and locomotive. But the comical 0–4–0, its 51-inch drivers turning resolutely, is up to the task. The A-1 begins to roll forward, through the erecting shop doors and out into the midday sun.

The A-1 and its entourage are headed for the testing shed to be met by an engineer and fireman. They climb up into the cab and begin checking valves and controls. An outside steam line is attached. Soon the locomotive begins to stir. Air compressors start their syncopated drumming. Feedwater pump and injector whir, drawing water from the tender to fill A-1's empty boiler. Then, with an explosive hiss, eight brake shoes set tightly to the drivers. A commotion up front; one of A-1's flags is missing from its holder on the pilot. Another is sent for. The sheen on the glossy, jet-black locomotive is blinding in the midday sun. Ben recognizes Petro up on the runningboard, checking each riveted seam for water, the sign of a boiler leak. He moves over below him to see how his riveting job stands the test of live steam. Petro turns to the boy waiting expectantly below. "Is okay, Benny. Good job."

Henry Koch arrives and begins setting up his cherry-wood 7 x 17 view camera, to take the A-1's official portrait. When Henry disappears under the focusing cloth and brings Will Woodard's new locomotive into sharp focus on the ground glass, more than the sun dazzles his eyes. Henry has taken portraits of hundreds of new locomotives, but this one, with its massive firebox and boiler, its straight lines and distinctive trailing truck, is unlike anything in his experience.

Up inside the cab, the fireman's attention is focused on the firebox. Pine kindling is lit and some coal shoveled in through the fire doors. Soon Ben can see a yellow glow flickering about the cab and in the grates under the firebox. Gray smoke begins drifting upward from the stack into the azure sky. Waves of heat ripple off the firebox. Clicks and raps and pings accompany the expansion of tons of hot steel plates and a thousand rivets.

The stoker begins to growl. The turbo-generator revs up to a high-pitched whine. A-1's headlight comes on bright against the daylight. Ben looks over to where Mr. Woodard is standing in the group of engineers and dignitaries. As if sensing the boy's attention, Will turns and looks over toward Ben. A smile flashes between them that needs no words. Will's round face breaks into a laugh; he reaches up out of the tight huddle around him and signals Ben a triumphant thumbs up.

Engineer Jim Boynton takes his seat at the window and looks out over the festive crowd. The bell up on the smokebox front begins its measured clanging, startling Ben and the others into rapt silence. A-1 exhales, and eight brake shoes relax their grip on the drivers. The power reverse sends a succession of movements through the links and levers of the valve gear. Little plumes of steam drift from the cylinder cocks.

The locomotive is alive. Two deafening, throaty blasts from A-1's whistle bring hurrahs and whoops from the crowd and a torrent of tears down Ben's cheeks. A second later the shrieks echo back from the tall glass and brick walls of the shops. A single puff of exhaust bursts from the stack, thrusting pistons and crossheads, hissing steam from the cylinders cocks, another burst from the stack, main rods gliding outward, *hissss* side rods turning bright steel-rimmed drivers. *Puffhisspuffhiss* rolling steel on steel. Ben walks alongside his locomotive barely keeping up. The A-1 slips away. He breaks into a run, faster and faster, the rods stroking at his shoulder, the wind in his face, the giant's heartbeat vibrating through his body, matching the cylinders breath for breath. The distance widens between them and A-1 rolls away from Ben out toward the test track. Ben slows to a walk and then stops. He watches after the locomotive, his hands thrust down into his overall pockets, until all he can see is the back of the tender getting smaller in the distance and great plumes of smoke dusting the sky.

"Well, we did it." Startled, Ben turns to find Mr. Woodard standing next to him. "She'll be out on the Boston & Albany next week for a series of test runs. I expect she'll do just fine. It's my guess you'll be building a lot of 2–8–4s in the years to come, Ben." Will reaches out for Ben's hand and holds it in a firm handshake. "My work's finished here for a while, so I'll be heading back to New York tomorrow. Good luck. And . . . I thought you'd like to have this." A mischievous smile comes over the engineer's face as his left hand comes from behind his back with a short wooden rod. Will unfurls the white flag and presents it to Ben. "I'm sure it won't be missed." He turns and hurries back toward the group awaiting him at the door of the erecting shop.

Ben looks down the track once more, but the A-1 is gone. So he heads back too, toward his dad waiting in the middle of the tracks. Ben's glad to see him. He's feeling sad, lonely, the way you feel when a buddy moves away. Together they walk to the machine shop. Ben is quiet and turned inward, and his dad understands. "I almost forgot," Ben snaps his fingers. "I've got a note from Joe that I have to take to Mr. Schnell. You go on. I'll catch up with you later."

◇

Mr. Schnell peers over the top of his glasses at the young man offering him a note. He pushes back from the big desk strewn with a chaos of blueprints and sketches, his face becoming more and more serious as he reads Joe's comments. The foreman removes his glasses, rubs his eyes wearily, looks up at Ben, and, with a touch of irritation in his voice, scolds the surprised helper. "You've really done it this time."

Ben is taken aback. It looks like he's in for a real lashing. And he doesn't know why. His mind races back over the events of the week. "I've been hearing all about your antics out there in the shop." Oh, oh, Ben says to himself, he's found out about the prank I pulled on Steph. How'd he hear about that? "And now, this note from Joe," the angry foreman says, slapping the wrinkled yellow scrap with the back of his hand. Ben has never seen Mr. Schnell this angry before. The foreman never shouts, but he's sure getting close to it now. "Yep, you've really gone and done it. How am I going to keep you on as my helper after this?" By now, Ben is trembling. "Joe says here that you are 'an especially industrious and earnest young man who I personally endorse and recommend for work in Department 12.'"

Charlie Schnell beams impishly as he reaches out to shake his helper's hand. "I'll talk with your dad and we'll get you started on a drill press Monday morning. And when you get good at that, we'll hunt up a lathe for you." With that, the foreman turns back to his work, chuckling. Ben, who is only just managing a smile, stammers out a "thank you" and heads for the door, for the first time aware of the sweat pouring down under his shirt.

"Oh, and by the way," Mr. Schnell calls out after him. Ben turns to see the graying head and bright eyes peering over their spectacles at him from around the big desk. "You'll be raised to 30 cents."

Acknowledgments

THE JOURNEY back to Lima in the 1920s, and The Locomotive out on south Main Street, took me into the lives of some remarkable people. Each gave me something special that enlivened the writing and drawing.

Walter Gray, Kevin Bunker, and Stephen Drew shared their knowledge of railroading and forgotten details of locomotive construction they had discovered under the flaking rust and paint of old engines restored for the California State Railroad Museum. Ellen Schwartz, the Museum's librarian, introduced me to the nuts-and-bolts textbooks that became my "correspondence" courses on locomotive building, boilermaking, machine shop, forge and foundry practice. Each time I visited the library, Ellen would have waiting for me piles of books and journals—always just what I needed.

When it was time to go to Lima, Ray Schuck welcomed me to the Allen County Museum and guided me through the largest collection of Lima Locomotive Works records—thousands of photographs, documents, and drawings. Ray and Mary Ann Schuck gave me a home while I worked at The Locomotive, and on summer evening walks showed me small-town Ohio as it might have been in Ben's day.

Ray introduced me to people with memories rich in detail and humor. Some of them stepped through the magic door and became characters in the story. Charles Schnell reminisced with me about his years as foreman. "All in all," he told me at the end, "it was a pretty interesting lifetime." Frances Park wrote me long, thoughtful letters about growing up in Lima. Later, when I met them, Fran and her husband Joe enchanted me for hours with their stories. It was Joe who told me about all the shenanigans that went on in the shop. But whenever he said what "those guys did," his grin gave him away.

On a day I'll never forget, Joe, John Keller, Ray Schuck and Bob Keller—who had taken a day off from his maple sugaring— walked me through The Locomotive, building by building, through doorways meant for giants, over acres of wood-block floors, into working places where they had spent their lives. On another day, John and I sat in the cab of Nickel Plate No. 779. They had retired together. He had been a head brakeman and spent many years right here in this seat behind the fireman. Now John recalled those years on the big 2–8–4 with the poignancy and poetry of an old railroad song. Lincoln Park was quiet and smelled of spring that morning, but I knew John was smelling coal smoke and was hearing his Berkshire working a long freight.

A lot of little pieces had to come together before the A-1 puzzle could be completed. Truson V. Buegel sent me the original erecting cards for the A-1 and its tender from his collection. Dr. Morton Malin and Gladys Breuer of The Franklin Institute researched the elusive Mr. Woodard for me. When I needed to know more about riveting and riveters, Nancy McBride helped with tools and techniques documented in the collections of the Virginia Museum of Transportation. And John H. White, Jr., was always there to answer my questions or offer approaches to those he could not answer. I used Jack's meticulous reconstructions of the *John Bull* and *Consolidation* as the basis for my drawings.

Over the years of writing and drawing, the author and his book enjoyed the attention of many friends. George Wotton helped me toward the clarity I wanted in my images. And Don Yoder took time off from his chores at Oakwood Farm to read and remark on Ben's story with his editor's keenness and poet's sensitivities. Robert Vogel's encouragement, criticism and engineering know-how have been more valuable than he could possibly know. Mariah was always there, too, at *The Shambles*, to read new chapters, to talk over coffee and Stephen's bread hot from the oven, about words, and writing and being writers.

SUPERPOWER

was set in Galliard by Meriden-Stinehour Press, Lunenburg, Vermont.
Designed by Matthew Carter and introduced in 1978 by the Mergenthaler
Linotype Company, Galliard is based on a type made by Robert Granjon
in the sixteenth century. It is the first type of its genre to be designed
exclusively for phototypesetting. A type of solid weight, Galliard
possesses the authentic sparkle that is lacking in the current
Garamonds. The italic is particularly felicitous and reaches
back to the feeling of the chancery style, from which
Claude Garamond in his italic had departed.

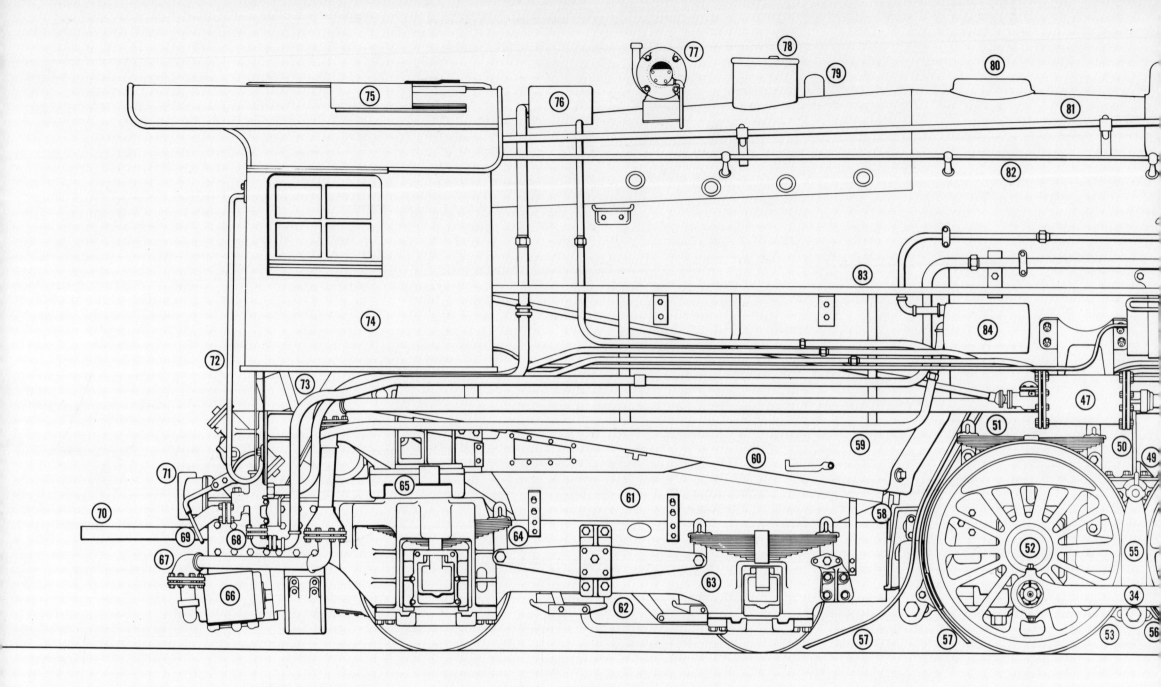